Georg Payrhuber

Hydrogen sulfide inhibits cell proliferation

Georg Payrhuber

Hydrogen sulfide inhibits cell proliferation

in C6 glioma and GH3 pituitary tumor cells

AV Akademikerverlag

Impressum / Imprint
Bibliografische Information der Deutschen Nationalbibliothek: Die Deutsche Nationalbibliothek verzeichnet diese Publikation in der Deutschen Nationalbibliografie; detaillierte bibliografische Daten sind im Internet über http://dnb.d-nb.de abrufbar.
Alle in diesem Buch genannten Marken und Produktnamen unterliegen warenzeichen-, marken- oder patentrechtlichem Schutz bzw. sind Warenzeichen oder eingetragene Warenzeichen der jeweiligen Inhaber. Die Wiedergabe von Marken, Produktnamen, Gebrauchsnamen, Handelsnamen, Warenbezeichnungen u.s.w. in diesem Werk berechtigt auch ohne besondere Kennzeichnung nicht zu der Annahme, dass solche Namen im Sinne der Warenzeichen- und Markenschutzgesetzgebung als frei zu betrachten wären und daher von jedermann benutzt werden dürften.

Bibliographic information published by the Deutsche Nationalbibliothek: The Deutsche Nationalbibliothek lists this publication in the Deutsche Nationalbibliografie; detailed bibliographic data are available in the Internet at http://dnb.d-nb.de.
Any brand names and product names mentioned in this book are subject to trademark, brand or patent protection and are trademarks or registered trademarks of their respective holders. The use of brand names, product names, common names, trade names, product descriptions etc. even without a particular marking in this work is in no way to be construed to mean that such names may be regarded as unrestricted in respect of trademark and brand protection legislation and could thus be used by anyone.

Coverbild / Cover image: www.ingimage.com

Verlag / Publisher:
AV Akademikerverlag
ist ein Imprint der / is a trademark of
OmniScriptum GmbH & Co. KG
Heinrich-Böcking-Str. 6-8, 66121 Saarbrücken, Deutschland / Germany
Email: info@akademikerverlag.de

Herstellung: siehe letzte Seite /
Printed at: see last page
ISBN: 978-3-639-79278-2

Content

List of abbreviations:

3MST : 3 - Mercaptopyruvat - sulfur - transferase

Akt : protein kinase B

AMPK : AMP (adenosine monophosphate) activated protein kinase

APP : amyloid precursor protein

ASK 1 : apoptosis signal-regulating kinase 1

ATP : adenosine tri-phosphate

bHLH : basic helix loop helix transcription factor

BK-channel : maxi calcium activated potassium channel

CAT : cysteine aminotransferase

CO : carbon monoxide

CBS : cystathionine beta-synthase

Cl⁻ : chloride

CSE : cystathionine γ-lyase

EC 50 : half maximal effective concentration

ER : endoplasmatic reticulum

ERK : extracellular-signal regulated kinase

GSK3β : glycogen synthase kinase 3 β

H_2O_2 : hydrogen peroxide

H_2S : hydrogen sulfide

Hes 1 : hairy and enhancer of split 1, transcription factor

HIF-1α : hypoxia induced factor 1 alpha

IC_{50} : half maximal inhibitory concentration

IQ : intelligence quotient

JNK : c-Jun N-terminal kinase

LC3B : ubiquitin-like light chain protein, part of the autophagosome

LDH : lactate dehydrogenase

LTM : long term memory

MAPK : mitogen activated protein kinase

MASH 1 : mammalian achaete scute homolog-1

mTOR : mechanistic target of rapamycin

NaHS : sodium hydrosulfate hydrate

Neuro D1 : neurogenic differentiation 1

Nng 1 : neurogenin 1

NO : nitric oxide

NSC : neuronal stem cells

ODC : ornithine decarboxylase

p38 : type of MAPK

Pen/Strep : penicillin/streptomycin

ROI : region of interest

TNF-α : tumor necrosis factor alpha

Introduction

Hydrogen sulfide

The gaseous compound hydrogen sulfide (H_2S) was first described in 1713. Subsequently papers were published describing the toxicological effects of H_2S, e.g. (Chernikov, 1952) and (Freidrich, 1946). Eventually in 1996 when Abe and Kimura found fairly high concentrations (50-160 µM) of H_2S in the rat brain (Abe and Kimura, 1996). They hypothesized that H_2S may play an important physiological role in the mammalian brain, similar to the gases nitric oxide (NO) and carbon monoxide (CO) which have been described as neuromodulators and gasotransmitters earlier (O'Dell et al., 1991), (Stevens and Wang, 1993). A more recent study shows that H_2S is present in much lower concentrations in the cerebrospinal fluid (CSF) of newborn pigs at a concentration of 561 ± 205 nM (Leffler et al., 2011).

Origin

To understand why a toxic gas such as H_2S may have found its way into our signal transduction systems one has to go back to prehistoric times. In the primordial atmosphere H_2S was present in high concentrations up to 6%. In addition to NO and CO, H_2S possibly also played a crucial role in the origin of life and the emergence of the mitochondria. At the beginning of life on earth oxygen was hardly present in the atmosphere. Thus, it is very likely that the first living beings have won their energy to a large percentage out of redox reactions with these three gases. In these reactions H_2S took over the role of electron donor (Olson et al., 2012).

Despite the enormous amount of time that has passed since then, this pathway was not lost entirely. In today's organisms these gases are found primarily as signal molecules called gasotransmitters. It is likely that H_2S has reached earth in different biochemical forms, like eruptions from vulcanoes, via comets, meteorites and interplanetary dust. H_2S was not only a reducing agent, but served the first organisms as a source of biocatalysts and structural elements in the form of iron-sulfur centers in proteins. Thus it is not surprising that 3.4 billion years old microfossils represent microbial ecosystems that are based on sulfur (Olson et al., 2012).

Production of H₂S in the body

The production of H_2S in the human body can be accomplished via three different pathways. The enzymes cystathionine beta-synthase (CBS), cystathionine γ-lyase (CSE) and cysteine aminotransferase (CAT), together with 3- Mercaptopyruvat - sulfur - transferase (3MST), use the proteinogenic amino acid cysteine as a source (see figure 1).

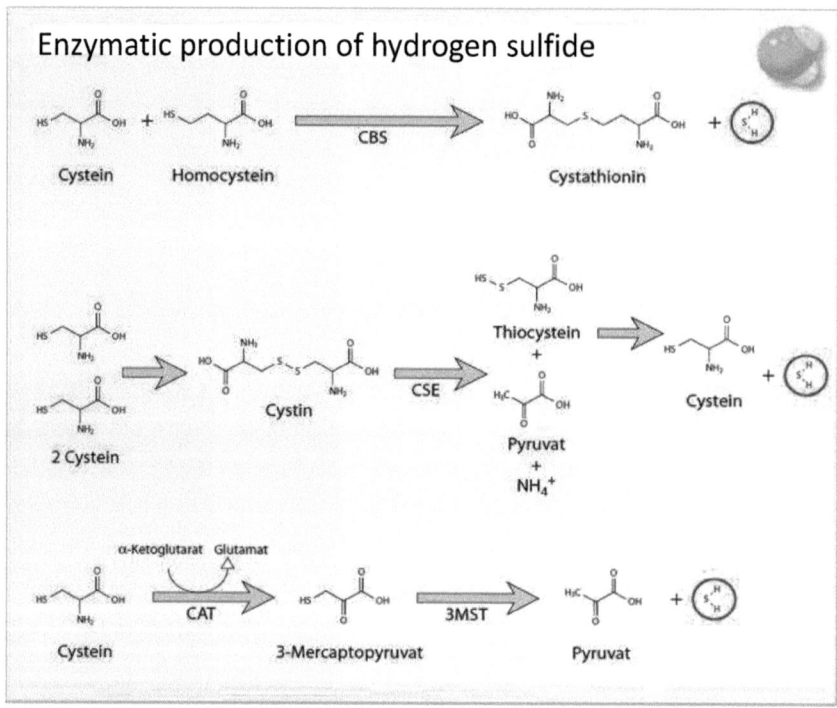

Enzymatic production of hydrogen sulfide

Cystein   Homocystein   CBS   Cystathionin

2 Cystein   Cystin   CSE   Thiocystein   Cystein   Pyruvat + NH₄⁺

α-Ketoglutarat Glutamat

Cystein   CAT   3-Mercaptopyruvat   3MST   Pyruvat

Figure 1: Production pathways of H₂S in the human body, edited from Hermann et al. (2011).

All three synthesis routes take place in the brain, where the highest concentration of H_2S was 9 µM (Kimura, 2010). Lower H_2S concentrations were found in liver (17 nM; Kimura, 2012), kidney (14 nM; Kimura, 2012) and blood (10 nM; Elsey et al., 2010). H_2S can either be released immediately after the production by the enzymes and thus be effective straight away, or it is stored in the form of acid - labile sulphur in iron-sulphur centers of enzymes, or as sulphur sulfane in the cytoplasm. If the H_2S production by CBS is limited, by a genetic mutation, it results in homocystinuria.

These changes lead to epilepsy or mental retardation in the affected patients (Elsey *et al.*, 2010) (Kimura, 2012).

If CBS, whose gene is located on chromosome 21, is overexpressed, which is the case in trisomy 21 for example, there will be elevated H_2S concentrations in the brain of these patients. This leads to mental underdevelopment, presumably because nerve cells die during the development of the brain due to the high concentration of H_2S. In addition, people with trisomy 21 have an increased Alzheimer's risk, as CBS occurs commonly in astrocytes and in the surrounding plaques. Furthermore, it was found that children with a higher than average intelligence quotient (IQ) often have less H_2S production in the brain, on the basis of other genetic mutations (Kimura, 2010).

Neuroprotection/Alzheimer´s disease
Alzheimer´s disease is one of the most common forms of neurodegenerative diseases these days, mainly caused by the accumulation of amyloid-β plaques in the brain. Sodium hydrosulfide hydrate (NaHS), a H_2S donor, can attenuate the decreasing viability of murine BV-2 microglia cells when incubated with amyloid-β peptides (Liu and Bian, 2010). This effect is concentration dependent with the highest impact at a NaHS concentration of 200 µM. The authors also found that a treatment with NaHS can revert the amyloid-β induced cell-cycle arrest and reduce the release of LDH, NO and TNF-α. All these effects are transduced via the p-38 (type of MAPK) and JNK-MAPKs (c-Jun N-terminal kinase mitogen activated protein kinases) pathways and require a NaHS treatment prior to the incubation with amyloid-β peptides.

A similar study (Giuliani *et al.*, 2013) shows that H_2S slows down the progression of Alzheimer´s disease. Different Alzheimer´s disease mouse models were treated with NaHS (0.25, 0.5 and 1 mg/kg) and H_2S rich spa-waters 3, 6 and 12 ml/kg i.p. (H_2S level in spa water were: 129 mg/l (HS^- 50 mg/l and not ionized H_2S 79 mg/l)). All models showed improvement in behavioral performance and reduced amyloid-β plaques in the hippocampus. According to the result of Liu et al. 2010 this study also reports a reduction of MAPKs, namely JNK, p38 and extracellular-signal regulated kinase (ERK). These kinases indicate a strong relation to inflammation and apoptosis.

Transgenic mice with increased expression of the amyloid precursor protein (APP) are often used as a model for Alzheimer's disease. He *et al.*, 2014 investigated the correlation of H_2S levels and the activity of the H_2S producing proteins, especially CBS, in mice with a progression of the disease. Intraperetonial application of 50 µmol/kg NaHS resulted in a reduction of caspase 3 expression, which is strongly linked to apoptosis. The administration of NaHS also resulted in increased learning and memory capacity, as indicated by the Morris water maze test. Treated mice had fewer difficulties finding the hidden platform. The screening for abundant amyloid plaques ensued that NaHS caused a shift from the plaque forming beta pathway to the non-plaque forming alpha pathway of APP cleavage. Thus the authors suggest a therapy with the application of exogenous H_2S donors to slow down the progression of Alzheimer's disease.

The anti-inflammatory effect and antioxidation properties of H_2S in this context were also described in earlier work by Hu *et al.*, 2007 and Tang *et al.*, 2008. A more recent study (Wang *et al.*, 2013a) puts L-cysteine in the role of the preferable H_2S donor. According to the authors the proteinogenic amino acid L-cysteine acts as a substrate for CBS and thereby promotes proliferation and differentiation of neural stem cells (NSC) in vitro. L-cysteine was added in a very low dose (1 µM) and the NSC were able to release significantly more H_2S into the surrounding culture media. Apparently the additional H_2S led to ERK phosphorylation which caused the gain in proliferation. Even though it was shown that this phosphorylation was caused by H_2S, the exact mechanism still remains unclear. The differentiation of NSCs is also not completely understood. L-cysteine promotes differentiation of NSCs by increasing expression of different transcription factors, e.g. mammalian achaete scute homolog-1 (Mash1), neurogenin 1 (Nng1) and neurogenic differentiation 1 (Neuro D1) (Wang *et al.*, 2013a).

Following these results, another publication shows very similar findings. Liu and colleagues (2014) discovered that NaHS increased phosphorylated ERK protein levels. Moreover, they found that some basic helix-loop-helix (bHLH) transcription factors which enhance neuronal differentiation are up-regulated, namely Mash1, Ngn1 and Neuro D2, whereas the repressor-type bHLH hairy and enhancer of split 1 (Hes 1) is down-regulated. They also showed that NaHS (0,1; 1; and 5 µM) increased proliferation of NSC in the dentate gyrus in mice subjected to hypoxia beforehand. In summary these studies indicate that NaHS improved the deficits of the cognitive impairment induced by hypoxia.

Effects of H₂S on proliferation

As mentioned above, H_2S has a positive effect on the proliferation of NSCs in the murine hippocampus (Liu *et al.*, 2014).Human colon cancer cells (SW480) exposed to 50, 100, 200 or 400 µM NaHS for 24, 48 or 72 hours showed improved proliferation, compared to the control group (Hong *et al.*, 2014). Pan *et al.* (2014) found that inhibition of CSE leads to a decreased proliferation in human hepatoma cell lines (HepG2 and PLC/PRF/5 cells). Recent studies also indicate that CBS could be the future target of anti-cancer drugs (as reviewed in Hellmich *et al.*, 2014). But there also exist reports on a decrease of cell viability after H_2S treatment (Perry et al., 2011; Yang *et al.*, 2004). Overexpressed CSE in HEK-293 cells for example results in inhibition of cell proliferation and DNA synthesis (Yang *et al.*, 2004). Primary human airway smooth muscle cells were treated with NaHS or GYY4137 (both 100 µM) for 48 or 72 hours (Perry et al., 2011). This resulted in decreased proliferation, compared to the control group. These different effects may depend on which cell types one is looking at and / or which concentrations are used.

In the study of Yang *et al.*, 2004 an elimination of all other CSE products like pyruvate and ammonium, left endogen H_2S as the trigger for the decreasing proliferation and DNA synthesis. As before ERK is involved in this mechanism, but in this case it is the activation of ERK that causes this effect. To verify the results exogenous H_2S was used (100 µM). H_2S was prepared as a 0,09 M stock solution from distilled water with pure H_2S gas. This way of H_2S application yielded the same results and thereby confirmed that H_2S is the reason for this outcome in HEK-293 cells (Yang *et al.*, 2004).

Similar results were obtained for an immortalized colon epithelial cell line (YAMC) and a colon cancer cell line (HT-29, SW1116, HCT116) (Wu *et al.*, 2012). Exposure of colon epithelial cells to H_2S led to inhibition of proliferation and G_1-phase cell cycle arrest. Increased formation of LC3B[+] autophagic vacuoles also indicate a strong link to autophagy (Wu *et al.*, 2012). (LC3B is a ubiquitin-like protein that is a constituent of the ATG8-conjugation system, one of two evolutionarily conserved phosphatidylethanolamine conjugation systems necessary for the formation of the autophagosome.) The effects were observed in normal as well as in cancerous colon epithelial cells (Wu et al., 2012). All these mechanisms are regulated by the AMPK/mTOR cascade, which plays a important role in regulating the energy balance, metabolism and cell growth (Wu et al., 2012;Xu et al., 2012).

Even though the proliferation of lymphocytes is stimulated by low concentrations (< 1mM) of NaHS, high concentrations (> 2 mM) of NaHS lead to inhibition of proliferation via AKT/GSK3 β signalling (Han et al., 2013). Cell viability and cell cycle arrest indicate the decreased proliferation rate. In contrast, CBS (cystathionine-β-synthase) was found to be overexpressed in tumor cells, compared to non-malignant cells (Szabo et al., 2013; Szabo and Hellmich, 2013). This leads to high concentrations of H_2S in the tissue. The authors found evidence that H_2S not only stimulates growth, migration and proliferation of colorectal cancer cells, but also promotes tumor angiogenesis.

Growth and development of rat glioma tumors after injection of NaHS were reported by Li et al. (2012). They injected C6 glioma cells into the brain of healthy rats. Afterwards a group of rats received NaHS intraperitoneally at a concentration of 35 nmol/g. Pathological analysis revealed that the C6 glioma was significantly larger when NaHS was administered. Additionally HIF-1α (hypoxia induced factor 1 alpha), a tumor marker protein, and angiogenesis were increased (Li et al., 2012).

Effects on behavior and learning
Recent studies show that H_2S has promising effects on behavior, and learning (e.g. Chen et al., 2013; Wang et al., 2013b). Chen and colleagues found that H_2S potentially yields antidepressant-like and anxiolytic-like effects in mice and rats (1.68 mg/kg NaHS was given intraperitoneally to the animals in this study). The forced swimming test, the elevated plus-maze test and the tail suspension test was used to determine antidepressant-like and anxiolytic-like effects of H_2S. The results show promising effects of H_2S, which could be used as a future antidepressant drug for people who experience severe side effects caused by the currently established

drugs. Hence, further research is needed in this field to understand the underlying mechanisms better.

Wang et al. (2013b) tested the effect of NaHS on hypoxia induced behavioral impairment in neonatal mice. To simulate hypoxia, the mice were treated with 5% oxygen for 120 minutes. Subsequently NaHS was given at 5.6 mg/kg for three days. Compared to the control group, NaHS significantly improved learning and memory performance and eased the delayed neuronal development of the mice. The effects appear to be mediated by an increased expression of brain derived neurotrophic factor (BDNF) and repression of NO- synthase (NOS) activity in the hippocampus.

Another study showed that H_2S can compromise the ability to learn and form LTMs in invertebrates (Rosenegger et al., 2004). Pond snails (Lymnaea stagnalis) were exposed to sodium sulfide (Na_2S), another H_2S donor. Considering that one third of the Na_2S forms H_2S, various concentrations ranging from 50 to 100 µM H_2S were used. Even though learning and formation of LTMs was observed at H_2S concentrations from 50 to 75 µM, it was considerably lower when compared to control conditions.

Cerebral ischemia can cause massive neuronal damage during the formation of the hippocampus. Wen et al., (2014) found in rats that NaHS (28 mmol/kg administered intraperetoneally for 7 days before ischemia) can improve the performance of rats in the Morris water maze. The animals had fewer difficulties in finding the platform and did spend less time in exploring the maze compared to untreated control animals after an ischemic stroke. These effects are mediated via an increased phosphorylation of Akt and a decreased phosphorylation of ASK 1 and JNK3, which both are involved in cancer, cardiovascular and neurodegenerative diseases.

Intracerebroventricular injection of homocysteine (0,2 to 2 µM) in rats causes learning and memory dysfunctions (Li et al., 2014). Furthermore CBS expression was lowered and thereby the endogenous H_2S concentrations. The authors also found that homocysteine upregulates marker proteins for ER stress, such as glucose-regulated protein 78 and cleaved caspase 12. Eventually the down regulation of endogenous produced H_2S on the one side and increased ER stress on the other hand results in learning and memory impairment.

CBS expression was also studied by Bruintjes *et al.*, (2014). They compared young mice (4 months old) to aged mice (24 or 28 months old). Immunostaining revealed that CBS is present in the whole brain, but notably in the hippocampus, cortex and amygdale of mice. Comparison of the three different age groups illustrates that there is no significant difference between them, in terms of CBS expression, although the older mice seemed to have a little increase in CBS expression. This indicates that maintenance and increase of CBS activity throughout age is important for the preservation of learning and memory in the hippocampus and other forebrain regions.

Intracerebroventricular injection of formaldehyde results in impairment of learning and memory and leads to increased apoptosis and lipidperoxidation in the rat hippocampus (Tang *et al.*, 2013). The decreased activity of CBS and the resulting decline in H_2S production were suggested to be the source of this disturbances caused by formaldehyde.

Interactions with channels
Ion channels are responsible for providing pathways to exchange any charge between the cytosol and the surrounding milieu. They are involved in the regulation of the cell osmolarity as well as electrical activity (such as action potentials), hormone release and cell cycle coordination, to name a few. Hence modulation of ion channels can effect a great variety of crucial mechanisms in an organism. All three gasotransmitters (CO, NO and H_2S) have been reported to affect different channels. Even though there still remain some uncertainties about the exact pathway and the definitive interaction partners/structures by which H_2S interacts with channels, there is no doubt about its significance to act as channel modulating agent (Hermann *et al.*, 2012).

One example of a channel which can be modified by H_2S is the maxi calcium activated potassium channel (BK-channel). With patch clamp whole cell recordings an increase of BK-outward currents was detected after the application of NaHS – an H_2S donor - on GH3 pituitary cells (Sitdikova *et al.*, 2010). Single channel recordings showed an increased open probability of BK-channels, in a concentration dependent manner. Different results were reported with HEK 293 cells (Telezhkin et al., 2009) where. NaHS inhibited BK-channel activity.

ATP-sensitive potassium channels in smooth muscle cells can be opened by H_2S. This eventually results in lower blood pressure and protection from ischemia and reperfusion injury of the heart (Ali et al., 2006; Zhao et al., 2001). Other channels modulated by H_2S include L-type Ca^{2+} channels (Gutiérrez-Martín et al., 2005), T-type Ca^{2+} channels (Matsunami et al., 2009), TRPV (Transient Receptor Potential Vanilloid) channels (Trevisani et al., 2005) and Cl^- channels (Kimura and Kimura, 2004). Furthermore, H_2S inhibits L-type Ca^{2+} channels in the heart, but activates them in neurons. It also protects neurons from oxytosis, oxidative stress induced cell death, via the activation of Cl^- channels (Peers et al., 2012; Tang et al., 2010).

Cell proliferation

Cell division, where a "mother cell" is dividing into two equal daughter cells is a main characteristic of cell proliferation.. To understand how proliferation works, one has to understand the cell cycle. All cells are permanently in one or the other stage of the cell cycle. Unlike bacteria or unicellular organisms, multicellular organisms usually do not pursuit exponential growth. The cell cycle is a strictly regulated cellular mechanism. Malfunction of this machinery can easily result in cancer (Löffler et al., 2007).

The cell cycle is usually divided into interphase, mitosis and cytokinesis or cytoplasmic division. The interphase consists of the G1, S and G2 phase. Beginning with the G1 phase, the cell is at rest and usually preparing to go into the S phase. A cell can also be permanently stay in G1 or G0 - as it is called then. Such cells, e.g. differentiated neurons or muscle cells, are unable to perform further cell division. If a cell is big enough, has a intact DNA and sufficient nutrients for a division, it can switch from G1 to the synthesis phase.

In the S phase the cell is reproducing its DNA and in addition RNA and proteins are produced in the cell. Once the DNA is duplicated, the cell proceeds into the G2 phase. If every requirement for the cell division is accomplished the cell can undergo mitosis. (Löffler et al., 2007).

Mitosis can be divided into five stages: prophase, prometaphase, metaphase, anaphase and telophase. During the prophase the chromatin of the cell starts to condensate. The resulting chromosomes consist of a pair of sister chromatids. They

13

are still attached to each other - the cell is in prometaphase. This phase is also characterized by the beginning formation of the mitotic spindle apparatus. In metaphase all chromosomes are aligned in the equator of the cell and each chromatid is attached to a microtubule from the spindle apparatus. Once the spindle apparatus starts to segregate the chromatid, the cell is in anaphase. The chromosomes move to opposite positions of the spindle. There the chromatid starts to decondense and form a new nucleus. The final telophase is characterized by the formation of a new nucleus and the beginning cytoplasmic division or cytokinesis. In the end two identical daughter cells are the result of the cell division (Alberts, 2002).

Regulation of the cell cycle is occurs via different cyclins and cyclin-dependent kinases. Throughout the cell cycle various checkpoints ensure that proliferation is controlled and only conducted if necessary. Cyclins are a group of proteins which cyclic vary in their abundance during the cell cycle. Additionally different growth factors drive the cell cycle. They are especially useful in cell culture. There they are added to the culture medium to make the cells to proliferate. Prominent examples are the platelet-derived growth factor (PDGF), the epidermal growth factor (EGF) or the fibroblast growth factor (FGF). PDGF is one of the first discovered growth factor with mitogenic properties, also known for its proliferation stimulating effects. Other growth factors can be stimulating or inhibiting, depending on the cell type, e.g. transforming growth factor β (TGF β). In epithelial cells TGH β was found to decline cell proliferation, whereas in cultured fibroblasts the proliferation is enhanced. (Clark *et al.*, 1998; Löffler *et al.*, 2007).

Polyamines

Putrescine, spermidine and spermine are the most common polyamines in prokaryotes and eukaryotes. They can modulate RNA as well as DNA, phospholipids and proteins (Igarashi and Kashiwagi, 2010). Proliferative stimuli increase the intracellular polyamine concentrations, indicating a strong link between polyamines and cell proliferation. Putrescine is synthesized from ornithine, spermidine from putrescine and spermine from spermidine. Ornithine decarboxylase (ODC), S-adenosylmethionine decarboxylase (SAMDC) and spermidine/spermine N1-acetyltransferase (SSAT) are the responsible enzymes for these reactions (see figure 2).

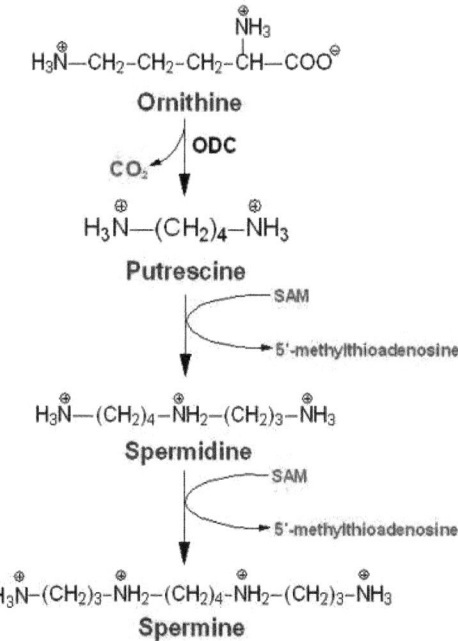

Figure 2: Synthesis of polyamines (ODC: Ornithine decarboxylase; SAM: S-adenosylmethlonlne decarboxylase), from Klng, 2014, http://themedicalbiochemistrypage.org/aminoacidderivatives.php.

ODC and/or SAMDC knockout-mice are unviable. This indicates the importance of polyamines for living cells. Polyamine concentrations are variable during the cell cycle (see figure 3.), with the highest concentrations in G1/S phase and G2/M phase shift (Weiger and Hermann, 2014; Yamashita *et al.*, 2013).

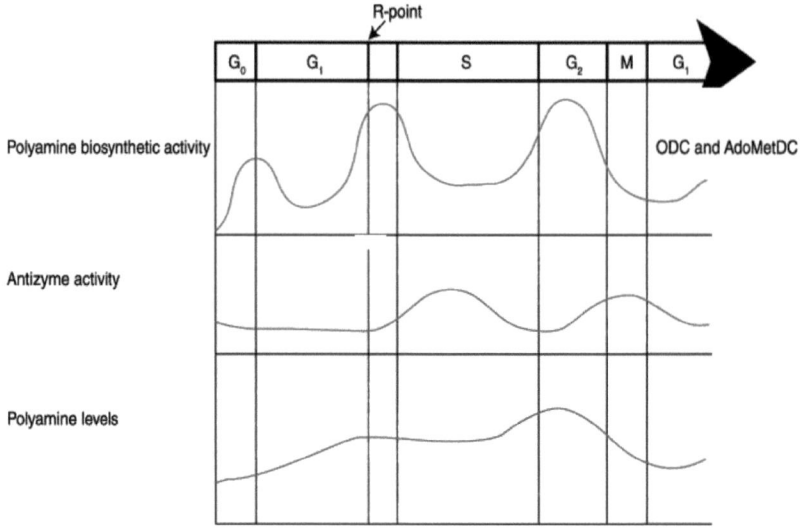

Figure 3: Timing of polyamine biosynthesis during the cell cycle, modified from: Alm and Oredsson, 2009.

Polyamines are charged and too big to just pass through the cell membrane. Therefore a specific transporter is required for polyamine uptake into the cell. In prokaryotes and yeast several polyamine transporters have been identified. Compared to these simple organisms the importance of polyamine transporters in mammalian cells is minor. Nevertheless, antizyme, a protein known for the degradation of ODC, was found to inhibit the uptake of polyamines. Additionally polyamines could be taken up via endocytosis (Igarashi and Kashiwagi, 2010).

Calcium signalling

Calcium (Ca^{2+}) is a most common second messenger in every living cell. It plays a role in gene expression, electrical activity, secretion, contraction and cellular metabolism. Intracellular Ca^{2+} levels can be changed either by releasing Ca^{2+} from intracellular compartments like the endoplasmic reticulum or by increasing the Ca^{2+} influx into the cell. The various Ca^{2+} signals can last for microseconds, like e.g during synaptic communication, or for seconds, e.g. in peristaltic contraction of gut muscle (Parekh and Penner, 1997).

Ca^{2+} has many different effector proteins, which either bind it directly or via an intermediate/adaptor protein. Thereby many different responses to a Ca^{2+} signal can be observed. Usually affinity, location and timing are the crucial factors that the Ca^{2+} response depends on.

H_2S and Ca^{2+} signalling

Both H_2S and Ca^{2+} are associated with proliferation, as mentioned above. Life imaging of Ca^{2+} changes in a cell might help to understand the connection between H_2S and Ca^{2+} signalling. For instance H_2S was found to increase intracellular Ca^{2+} concentrations and to induce cell migration in tumor-derived endothelial cells (Pupo et al., 2011). The authors used human breast carcinoma cells (B-TECs) and "normal" human microvascular endothelial cells (HMVECs). Results show that Ca^{2+} signalling was evoked in both cell types by H_2S. Extracellular application of EGTA (ethylene glycol tetraacetic acid), a Ca^{2+} chelator, led to inhibition of Ca^{2+} entry from the extracellular medium and abolished Ca^{2+} signals. However, the responses are very different between TECs and HMVECs. TECs seem to have a higher sensitivity to NaHS. Zhao et al. (2013) found that H_2S significantly promotes proliferation in human induced pluripotent stem cell-derived mesenchymal stromal cells. Exogenous NaHS prevented apoptosis and suppressed BK currents (Zhao et al., 2013). The authors obtained similar results, when they used Paxilline as BK channel blocker.

In a recent study cultured rat cardiomyoblasts (H9c2) were treated with 10 µM NaHS (Avanzato et al., 2014). This resulted in a decreased intracellular Ca^{2+}concentration. Control experiments with Nifedipine (10 µM), a known L-type Ca^{2+} channel inhibitor, confirmed that H_2S blocks L-type Ca^{2+} channels. NaHS also prevented the cells from H_2O_2 (hydrogen peroxide) induced cell death.

Sekiguchi et al. (2014) explored the effet of H_2S on T-type Ca^{2+} channels. They used cav 3.2. (a T-type Ca^{2+} channel) transfected HEK 293 cells. Dl-propargylglycine (PPG), a CSE inhibitor, significantly decreased T-currents in these cells. 1,5 mM NaHS reverted this effect and T-currents were increased above control level (Sekiguchi et al. 2014).

In summary the above mentioned studies (Avanzato *et al.*, 2014, Pupo *et al.*, 2011; Sekiguchi *et al.* 2014 and Zhao *et al.*, 2013) show that H_2S modulates different types of Ca^{2+} channels and hence appears crucial for Ca^{2+} signalling and proliferation.

H_2S donors

Different H_2S donors are available. The most straight forward donor is the gas itself. It can be stored in pressure tanks and used for fumigation of animals or bubbling of solutions. For experiments on cellular level pharmacological donors of H_2S are more suitable. ACS49 (CTG Pharma) is a donor which releases H_2S in a ratio of approximately 1:1 (Pupo *et al.*, 2011). Other donors are GYY4137, which is a very slow releasing H_2S donor, diallyl trisulfide, ATB-343 and NaHS. GYY4137 for instance was found to release much less H_2S than NaHS. Diallyl trisulfide releases about 1/10 of its concentration as H_2S (Predmore *et al.*, 2012). Approximately 1/40 of GYY4137 is released as H_2S, compared to 1/3 from NaHS (Lee *et al.*, 2011; Reiffenstein *et al.*, 1992). NaHS dissociates to Na^+ and HS^- in solution. HS^- associates with H^+ to produce H_2S. In a solution of pH 7.4 approximately one-third of NaHS exists as H_2S and the remaining two-thirds are present as HS^- according to the literature published at the time of the preparation of this thesis. Recent recalculation of the H_2S concentration derived from NaHS taking pH, temperature as well as salinity and evaporation into account estimate a remaining H_2S concentration of 11-13% in the experimental solution (Hermann et al., personal communication 2014).

Aims

One goal of my work with H_2S was to explore the influence of the gas on the proliferation rate of C6 glioma and GH3 pituitary tumor cells. The concentration dependence of this effect was particularly important. H_2S was found to decrease cell proliferation of both cell types.

A further hypothesis was that polyamines which are crucial for cells to proceed through the cycle may overcome proliferation restricting effects of H_2S. Therefore putrescine, spermidine, and spermine were added to the culture medium – however, without any effect.

Another goal was to uncover if H_2S changes Ca^{2+} concentrations in cells and thereby controls apoptosis and/or proliferation. Ca^{2+} imaging was used to accomplish this objective.

Material and Methods

Apparatus and Chemicals

All used apparatus, computer programs and chemicals are listed in Table 1-3.

Apparatus	Manufacturer/Model
Analytical balance	AND/ ER-182 A
Calcium imaging recording chamber	Leica RC-20
Calcium imaging system	Till Photonics (Oligochrome)
Cell counting chamber	Neubauer improved Zählkammer
Cell counting device	Millipore/Scepter with 60 µl sensors
Centrifuge	Sigma/3-16 K
Incubator	Thermo/Hera/cell 150
Laboratory scale	Sartorius
Laminar flow cabinet	Thermo/Hera Safe/KS 18
Microscope	Leica (DM IRB)
PC	Dell, Microsoft Windows 7
	ALA Scientific instruments (ALA-VM4)
Pipetting aid (electronic)	Accujet

Table 1: Equipment.

Methods	Program
Calcium imaging data analysis	FEI software (Version: 2.4.0.15); FEI Munich
Calculations	Microsoft, Excel 2007
Citation	Citavi 4, by Swiss Academic Software GmbH
Data analysis	Scepter Software Pro 2.1, by Millipore
Graphics	Graphpad Inc., Prism 5
Statistics	IBM SPSS Statistics 20

Table 2: Methods.

Chemicals	Manufacturer
Aqua bidest	provided by the department
Calcium Chloride	Sigma-Aldrich (C4901)
D-(+)-Glucose	Sigma-Aldrich (G8270)
Di-sodium-hydrogen-phosphate	Merck (6346.0500)
DMSO	Sigma-Aldrich (D2650)
Ethanol (70%) for cleaning	Merck
Fetal bovine serum	Gibco (Lot: 41G3681 K)
Fura-2 AM	Invitrogen (F1221)
Hepes [2-(4-(2-Hydroxyethyl)- 1-piperazinyl)-ethansulfonsäure)]	Sigma-Aldrich (H3375)
Horse serum	Gibco (Lot: 679427 D)
Magnesium Chloride	Sigma-Aldrich (M9272)
Medium, MEM Eagle	Sigma (M4655)
Medium, Nutrient Mixture F-10 Ham	Gibco (REF 31550-023)
Medium, Optimem	Gibco (REF 51985-026)
Penicillin-Streptomycin	Gibco BRL (Cat. No. 15140-122)
Pluronic F-127	Sigma (P-2443)
Potassium chloride	Sigma-Aldrich (P9541)
Potassium-Dihydrogen-phosphate	Riedel-de Haën (04243)
Putrescine	Sigma-Aldrich (P5780)
Sodium chloride	Sigma-Aldrich (S7653)
Sodium hydrosulfide hydrate	Sigma-Aldrich (161527)
Spermidine	Sigma-Aldrich (85570)
Spermine	Sigma-Aldrich (S3256)
Trypan blue solution cell culture tested	Sigma-Aldrich (T8154)
Trypsin/EDTA (0,05%)	Gibco (REF 25300-062)

Table 3: Chemicals.

Cells

Immortalized cell lines were used for all experiments. Compared to in vivo experiments these in vitro studies have significant advantages such as unlimited amounts of cells, easy handling and going back to the start passage or store them for a long period of time, thanks to cryopreservation.

C6 rat glioma and GH3 cells derived from the rat pituitary gland, were utilized in the present study. GH3 as well as C6 cells were used for dose response experiments GH3 cells were used for Ca^{2+} imaging.

Both cell-types were cultivated in an incubator with 90% humidity, 37°C and 5 % CO_2.

Cell culture

C6 glia cells were used from internally counted passage 9 to passage 20 and GH3 pituitary cells from internally counted passage 19 to passage 45. For C6 cells a medium was used consisting of 450 ml Nutrient Mixture F10 Ham supplemented with 44 ml FBS and 4,4 ml Pen/Strep. GH3 cells were cultured in a culture medium consisting of 445 ml MEM Eagle supplemented with 35 ml FBS (fetal bovine serum), 15 ml HS (horse serum) and 4,4 ml Pen/Strep. Pen/Strep was necessary in all experiments with NaHS. Due to the toxicity of NaHS, the medium had to be changed under the fume cupboard. The septic conditions in this area, in contrast to the conditions which were present in the routine sterile hood of the cell culture lab, made the use of antibiotics necessary.

All cells were cultured for routine propagation in cell culture flasks containing 6 ml of the appropriate cell culture medium. I exchanged the medium of the cells every third day. Splitting was necessary if a high density of cells was reached or if the whole base of the cell culture flask was occupied by cells. For splitting the medium was aspirated using a 2 ml trypsine-EDTA solution to detach the cells from the cell culture flask. Cells were transferred into a centrifuge tube, containing 8 ml of regular medium. After centrifuging for 5 minutes at 200 g at room temperature, the supernatant was discarded. I added 1 ml regular medium to the pellet and resuspended it with a pipette. Approximately one drop of the resulting cell suspension was transferred into a new cell culture flask filled with 6 ml new medium. After this process, the passage count was increased by one number.

Hydrogen sulfide donor

Sodium hydrosulfide hydrate (NaHS) was used as H_2S donor in all experiments. Since H_2S evaporates all NaHS solutions for the Ca^{2+} imaging experiments were prepared shortly prior to the experiment and were in use no longer than 20 minutes after NaHS preparation.

Experiments

Cell counting

To determine the exact cell number, the Scepter™ device from Millipore was employed. C6 or GH3 cells were seeded in a concentration of 2.5×10^4 cells/ml in 35 mm cell culture dishes. Every dish was filled with 2 ml cell suspension.

The concentration was determined by using the Neubauer cell counting chamber and then confirmed by the Scepter measurements. To do so 10 µl of the cell suspension were mixed with 10 µl Trypanblue and 10 µl transferred into the counting chamber. Six central fields in the Neubauer cell counting chamber were counted, an arithmetic mean was calculated and via multiplication by 50×10^4 the cell count was obtained. This quotient is derived from the volume of the central square = 1 mm^2 * 0.1 mm = 0,1 mm^3 = $1*10^{-4}$ ml and the dilution factor 5*10=50.

Once the cell number was known, the appropriate volume of the cell suspension was transferred into a centrifugal tube containing 13 ml medium. After mixing 12 ml were distributed into 6 cell culture dishes and 1 ml was transferred into a small glass beaker. Here the cell number was determined by the Scepter device. A 60 µl tip was attached to the device and the instructions of the producer were followed.

After seeding the cells were allowed for 3-5 hours to settle on the plastic dish. Then the medium was changed and experimental media containing NaHS or polyamines were added to half of the culture dishes. NaHS concentrations varied between 10 µM and 3000 µM (effective approx. 1 - 300 µM), whereas polyamines were added always at a concentration of 3 and 30 µM. If NaHS was used, all cell culture dishes were sealed with Parafilm, to prevent evaporation of H_2S. To guarantee sterile environment, in all experiments with NaHS Pen/Strep (1%) was added to the medium as described above. The H_2S donor had to be added to the cells under the laboratory hood due to its toxicity. Lacking aseptic conditions, the lab hood was a source of possible contaminations. Pen/Strep was added also to the control cells in the same concentration.

Experiments with 24 hours H₂S incubation

Cells were incubated at 37°C, 5 % CO_2 and medium containing NaHS for 24 hours. Then the medium was changed to a fresh medium without NaHS and the cells were incubated for another 24 hours under control conditions. Afterwards the cells were counted with the Scepter.

Experiments with 48 hours H₂S incubation

Cells were incubated at 37°C, 5 % CO_2 and medium containing NaHS or polyamines for 24 hours. Then the medium was changed to a fresh medium containing the same amount of NaHS or polyamines and the cells were incubated for another 24 hours. Afterwards the cells were counted with the Scepter.

Counting of cells

To detach and prepare cells for counting 0.05 % Trypsine-EDTA was used. First the experimental medium was aspirated, then 2 ml Trypsine-EDTA were transferred into every cell culture dish. To ensure that all cells were detached from the cell culture dish, the dishes were gently knocked by hand at the bottom. Next, cells were transferred into a 15 ml centrifuge tube containing 8 ml control medium. After centrifugation at 200 g at room temperature for 5 minutes the cells were resuspended in 1 ml medium. Finally they were counted with the Millipore Scepter device as described earlier.

In each experimental condition three control and three experimental cell culture dishes were used. The arithmetic mean of the cell number counted in each individual dish was calculated from each group.

The data obtained with the Scepter cell counting device were uploaded and stored on a PC. The data were then analyzed using the Scepter software Pro 2.1 (Millipore, USA) and eventually transferred into Microsoft Excel for further calculations. Histograms representing the distribution of cells as a function of diameter were generated and analyzed (figure 4):

The large peak in the graph of figure 4 marked with M1 represents the amount and distribution of single living cells. To the left of M1 the cell diameters are smaller compared to the diameters of normal cells. Therefore, one finds in this area of the histogram cells that are most likely passing into apoptosis (marked M2). On the right side of the histogram cell aggregates marked M3 are located. M1 contains single individual cells. Therefore, markers were placed in the valley between the individual and apoptotic cells (at 8 microns cell diameter) and on the other side at about 24 microns cell diameter, which is about equal to twice the mean cell diameter and thus should include all single cells. The remnants on the left and right side were defined as M2 and M3 (see figure 4.). For the determination of the cell count only M1 was used.

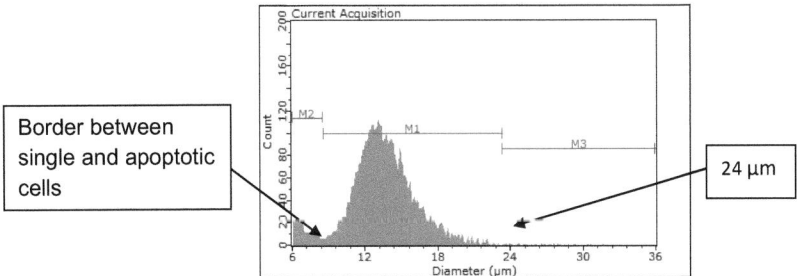

Figure 4: Cell counting data from the Scepter, x axis represents cell diameter in µm, the y axis represents the number of cells counted.

Using Excel, the cells seeded were subtracted from the cell number which was measured at the end of the experiment to determine the net proliferation. :

$$\text{Net proliferation} = \text{cell count after experiment} - \text{number of seeded cells}$$

GraphPad Prism 5 was used for statistical analysis and presentation of the data.

Ca^{2+} imaging

Fura-2-AM was used for Ca^{2+} imaging. In this compound the dye Fura-2 is linked to a acetoxymethyl ester (AM). This ester is responsible for the cell membrane permeability of the dye. Once inside the cell, esterases cleave of the acetoxymethyl

25

ester and the dye is freely available in the cytosol. All dyes are very sensitive to light; therefore one has to make sure to work in a darkened area (Parys *et al.*, 2014).

For Ca^{2+} imaging a ratiometric measurement technique was used (see below). Glass cover slips with a diameter of 12 mm were coated with poly-d-lysine to guarantee the adherence of the cells on the glass. Three cover slips were placed in a 35 mm cell culture dish. In each dish GH3 cells were seeded at $2,5*10^4$ cells/ml. The culture dishes were incubated at 37°C, 5 % CO_2 and 90% humidity in an incubator for at least three days. Cells were used for Ca^{2+} imaging at day three until day five after seeding.

A Ca^{2+} imaging buffer was prepared according to table 4.

pH 7,4	mM	g/l
NaCl	140	8,182
KCl	3	0,2235
CaCl$_2$	4	0,444
MgCl$_2$	1	0,203
HEPES	10	2,383
Glucose	10	1,802

Table 4: Buffer for Ca^{2+} imaging.

A buffer containing 15 mM KCl which was used to depolarize cells at the end of the experiment in order to increase intracellular Ca^{2+} was prepared containing 200 ml Ca^{2+} imaging buffer supplemented with 0,178 g KCl (12 mM). The cells were loaded with Fura-2-AM following this protocol:

0.01g Pluronic (to better disperse the Fura-2-AM) was dissolved in 50 µl DMSO, in a 37°C bath. 50 µg Fura-2-AM was added to 50 µl of a DMSO+Pluronic solution. Cells were washed with 1 ml Ca^{2+} imaging buffer (table 4) once, then 1 ml buffer (table 4) plus 5 µl Fura-2-AM was added. Then the cells were incubated for 50 minutes at 37°C in the dark. Afterwards the cells were washed twice with buffer (table 4) and incubated with buffer (table 4) for another 30 minutes at 37°C in the dark.

One cover slip was then placed in the recording chamber (Leica, RC-20) filled with control buffer solution (table 4). A perfusion system from ALA Scientific Instruments (ALA-VM4) was used to change solutions.

Once the recording chamber on the microscope was in place, a suitable group of cells for imaging was identified.

The light source for my Ca^{2+} imaging experiments was a 150 watt xenon lamp (Oligochrome by Till Photonics). Fura-2 excitation is accomplished at 387 nm and 340 nm (ratiometric measurement). The detection wavelength is 510 nm. (Simpson,2011). The online acquisition software (Version: 2.4.0.15) from FEI Munich at the Ca^{2+} imaging device from Till Photonics automatically calculates the calcium ratio for different regions of interest (ROIs), which were determined before. This is done via the equation:

$$R = F_{340}/F_{380}$$

R = fluorescence intensity ratio

A crucial step is the placement of a background ROI (figure 5). This is necessary for the correct calculation of the Ca^{2+} ratio, as the background has to be subtracted from the original signal to obtain the proper Ca^{2+} signal.

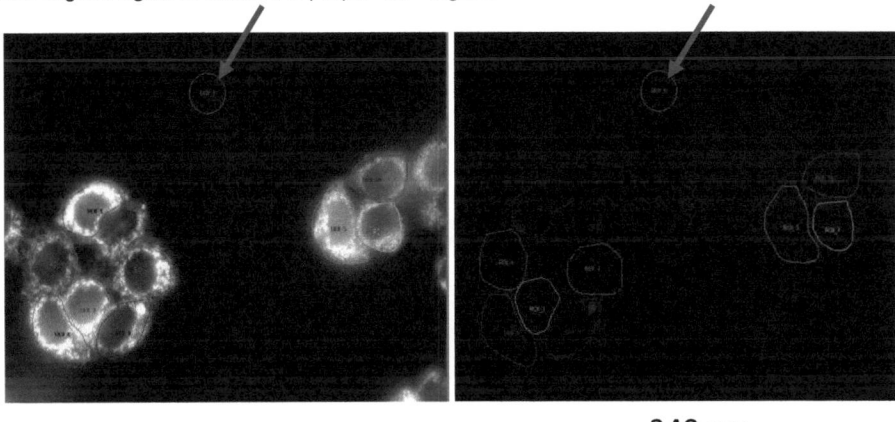

378 nm   340 nm

Figure 5: Calcium imaging with Fura-2 AM, original data: Comparison of cell fluorescence at 378 nm and 340 nm, placement of ROIs (region of interest), arrow points to the background ROI in this example (pictures were taken with online acquisition software (Version 2.4.0.15, from Till photonics).

The raw data from imaging experiments were exported as an excel file and further statistically analysed. In these excel files the response of each ROI is listed, sample intervals were 1.409 seconds. To calculate the response for control, NaHS, control 2, KCl and washout an average, for every condition, was computed. A total of 349 different cells was recorded and analysed.

Statistical analysis

Raw data were first analysed with Microsoft Excel. All statistical tests were conducted with SPSS. Graphical figures were created with Graphpad Prism 6.

The raw data in the cell counting experiments with NaHS and polyamines were standardised and expressed in percent of control which was set to 100%. This allows for a better comparability of the data.

In Ca^{2+} imaging experiments the generated data, the ratio of the response at 340 nm and 378 nm, were first logarithmised and then statistically analysed. This step guarantees a normal distribution of the data and to the subsequent of use parametric tests.

Results

C6 net proliferation (48 and 24 hours incubation with NaHS)

C6 cells were counted at the end of the incubation time, the net proliferation computed and standardized (control =100%) and an arithmetic mean was calculated for each NaHS concentration. The number of repetitions (N) for each experimental condition was between 2 and 4.

C6 cells 48 hours incubation with NaHS:

Figures 6 and 7 show the cell Scepter counting data from C6 cells after 48 hours with or without 1000 µM NaHS. Both graphs were taken from the same experiment and demonstrate that 1000 µM NaHS reduced cell proliferation.

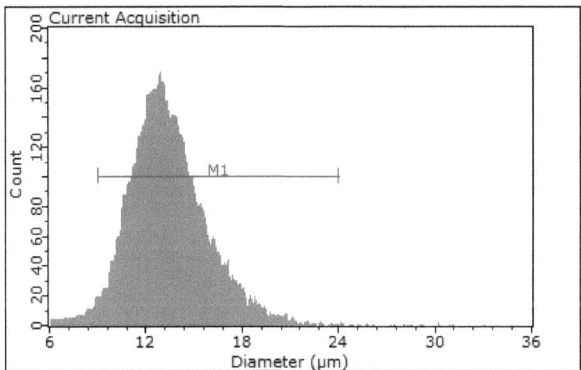

Figure 6: Cell counting data for C6 cells 48 hours in control medium, x axis represents cell diameter in µm, the y axis represents the number of cells counted. M1 represents the data area used for calculating the total cell number.

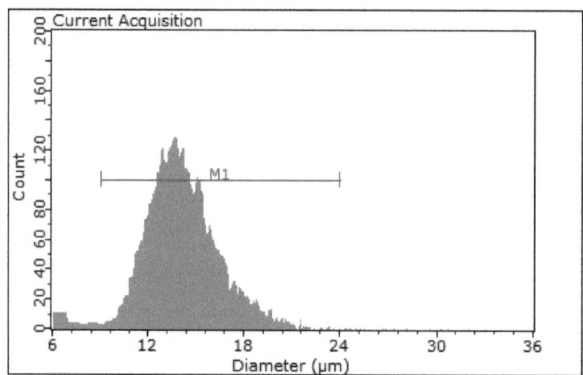

Figure 7: Cell counting data for C6 cells incubated with 1000 µM NaHS for 48 hours, x axis represents cell diameter in µm, the y axis represents the number of cells counted. M1 represents the data area used for calculating the total cell number.

NaHS (µM)	mean of net proliferation (%)	SD	N
0	100	-	3
3	104.850	3.750	2
10	99.460	11.090	3
30	119.960	12.040	2
100	91.610	6.870	3
300	75.860	13.470	3
1000	52.150	13.610	4
3000	2.800	9.960	2

Table 5: Mean of net proliferation with control set to 100%, C6 cells, treated for 48 hours with various concentrations of NaHS (SD = standard deviation; N = sample size).

Table 5 shows net proliferation of C6 cells treated with various concentrations of NaHS for 48 hours.. It was found that NaHS decreased the net proliferation in a dose dependent manner.

Data were transferred to GraphPad Prism software and dose response curves (figures 8,11,12,15 and 18) were created, using the Hill equation to fit the data and to determine the Hill coefficient as well as the half maximal effective concentration (EC50):

$$Y = 1 - \left(\frac{\left[x^{Hill\ coefficient} \right]}{\left[x^{Hill\ coefficient} + EC50^{Hill\ coefficient} \right]} \right)$$

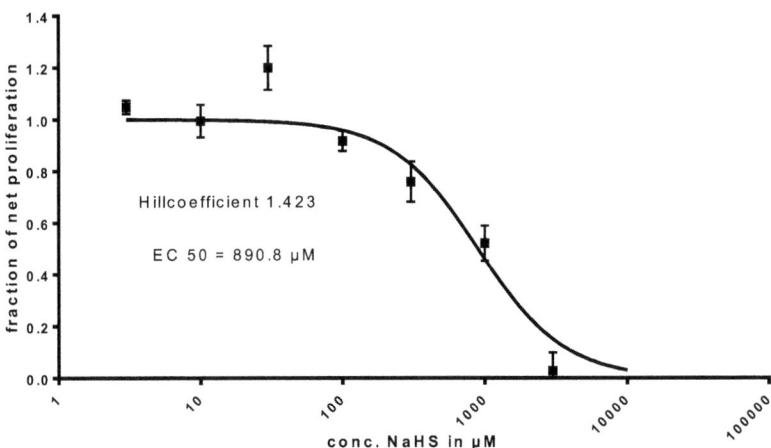

C 6 48h in NaHS

conc. NaHS in µM

Hillcoefficient 1.423

EC 50 = 890.8 µM

Figure 8: Dose response curve, C6 cells treated for 48 hours with NaHS.

The data from table 5 were used to create figure 8. The EC50 value represents the half maximal concentration, where cell net proliferation is at 50 % of the control level. In this experimental setting an EC50 of 890.8 µM NaHS indicates a reduction of the net proliferation rate by 50 %, when incubated with NaHS at this concentration for 48 hours. The Hill coefficient reflects the steepness of the dose response curve. It is a measure of the cooperative interaction between an enzyme and a substrate or in a wider sense between a target and its effector. A value greater than 1 is considered a positive cooperation, whereas a value less than 1 is an indicator for a negative cooperation. For this experiment it looks like H_2S cooperatively interacts with a target in the C6 cells, when incubated for 48 hours with NaHS, as indicated by the Hill coefficient of 1.423. It is interesting to note that 3 µM and 30 µM NaHS have a positive effect on the net proliferation of C6 cells. In both cases the proliferation is increased non significantly. All other concentrations of NaHS tested did not enhance the net proliferation above the control level of 100 % but caused a decrease of cell proliferation. 10 µM NaHS resulted in a net proliferation of 99,46 % which is close to 100 % and does not seem to have any effect on cell proliferation

31

C6 cells 24 hours incubation with NaHS and 24 hours recovery in NaHS free medium:

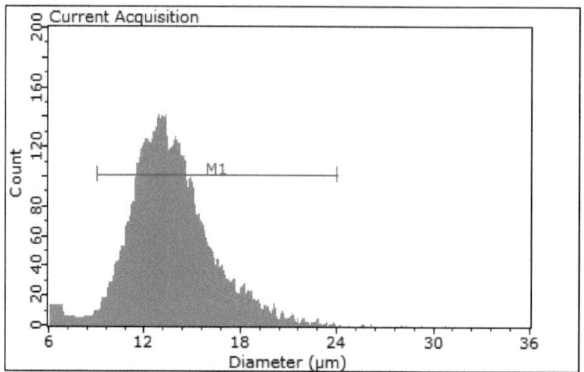

Figure 9: Cell counting data for C6 cells 48 hours incubated with control medium, x axis represents cell diameter in μm, the y axis represents the number of cells counted. M1 represents the data area used for calculating the total cell number.

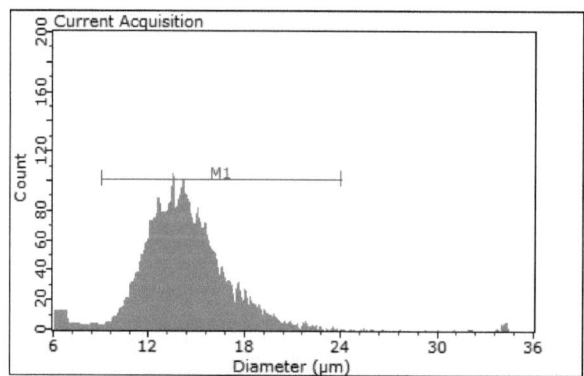

Figure 10: Cell counting data for C6 cells and 1000 μM NaHS for 24 hours and 24 hours recovery in NaHS free control medium, x axis represents cell diameter in μm, the y axis represents the number of cells counted. M1 represents the data area used for calculating the total cell number.

Figures 9 and 10 show the raw cell counting data from C6 cells after a 24 hour period of incubation with or without 1000 μM NaHS and a subsequent 24 hours recovery phase in NaHS free control medium, as obtained with the Scepter counting device.

32

Both graphs were taken from the same experiment. Around 6 µm one can see a small peak in both graphs. This peak represents the dead and probably apoptotic cells. All other as viable considered cells have a diameter from 9 µm to 24 µm.

NaHS (µM)	mean of net proliferation (%)	SD	N
0	100	-	3
3	96.610	5.920	3
10	102.680	6.668	3
30	90.000	13.757	3
100	90.460	6.471	4
300	85.300	1.970	3
1000	60.160	6.542	4
3000	55.720	14.946	3

Table 6: Dose response: Mean of net proliferation with control set to 100%, C6 cells, treated for 24 hours with various concentrations of NaHS followed by an additional 24 hours recovery phase.

In table 6 the results of an experiment are presented where C6 cells were treated with various concentrations of NaHS for 24 hours followed by a 24 hour recovery phase in NaHS free control medium. NaHS decreased net proliferation in a dose dependent manner.

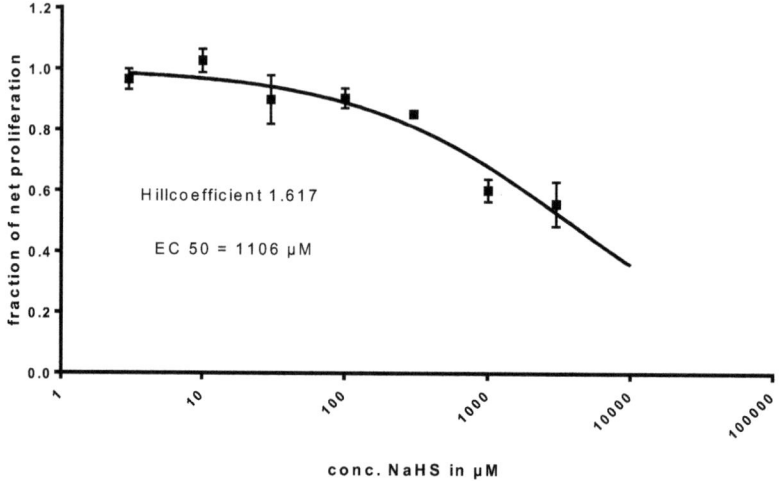

Figure 11: Dose response curve for C6 cells, treated for 24 hours with NaHS followed by a 24 hours recovery phase in NaHS free control media.

The data from table 6 were used to create figure 11. The EC 50 value of 3650 µM NaHS means that the net proliferation rate of C6 cells is reduced by 50 %, when incubated with NaHS for 24 hours at that concentration followed by a recovery phase of 24 hours in a NaHS free control medium. The Hill coefficient of 0.5821 indicates no cooperative interaction between H_2S and C6 cell proliferation. At the concentration of 10 µM NaHS a small but enhanced net proliferation was observed (102,68%), compared to the control level of 100 %. This increase in net proliferation was not significant, when compared to control. All other NaHS concentrations decrease the net proliferation despite an 24 hour recovery phase.

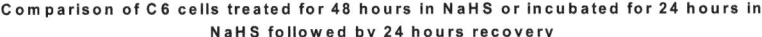

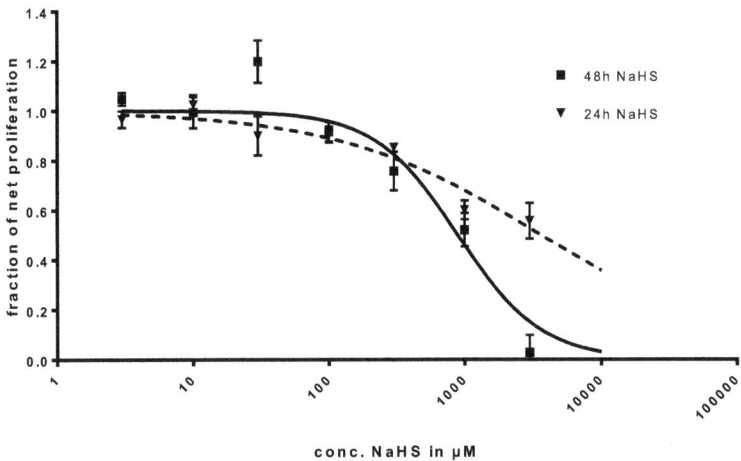

Figure 12: Comparison of net proliferation dose response curves for C6 cells treated either for 48h with NaHS or for 24 hours followed by a 24 hour recovery period in control medium.

To enable a better comparison between the results of the experiment with C6 cells treated 48 hours with NaHS and the experiment with C6 cells treated 24 hours with NaHS and a subsequent recovery phase of 24 hours in NaHS free control medium, figure 12 was created. One can see that there is a great similarity between the two curves in the lower concentration range (from 3 to 1000 μM NaHS). But starting with 3000 μM NaHS the two data sets drift apart indicating recovery of cells from NaHS treatment.

GH3 net proliferation (48 and 24 hours incubation with NaHS)

GH3 cells were counted at the end of the incubation time and an arithmetic mean was calculated for each NaHS concentration. The sample size was two or three. For 0 μM NaHS = control, the net proliferation was set to 100 %.

GH3 cells 48 hours incubation with NaHS

Figures 13 and 14 show the cell counting data from GH3 cells after 48 hours with or without 1000 μM NaHS, as obtained with the Scepter cell counting device. Both graphs were taken from the same experiment. One can see that the peak in figure 9 is much smaller than in figure 8 indicating a lower number of cells.

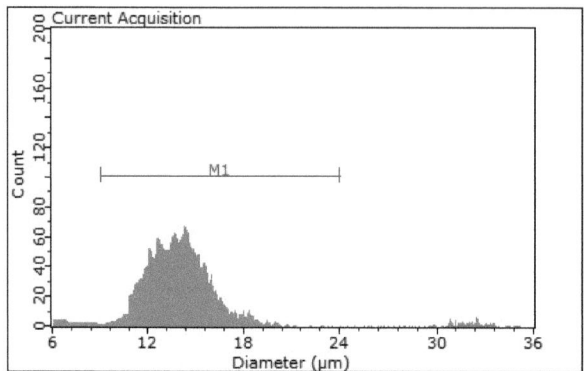

Figure 13: Cell counting data for GH3 cells incubated for 48 hours in control medium, x axis represents cell diameter in μm, the y axis represents the number of cells counted. M1 indicates the data area used for calculating the total cell number.

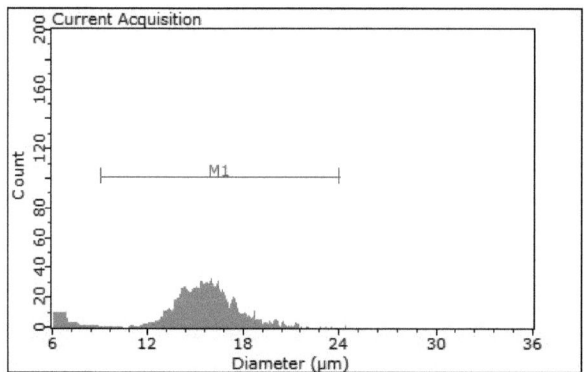

Figure 14: Cell counting data device for GH3 cells incubated in 1000 μM NaHS for 48 hours, x axis represents cell diameter in μm, the y axis represents the number of cells counted. M1 indicates the data area used for calculating the total cell number.

NaHS (µM)	mean of net proliferation (%)	SD	N
0	100	-	3
1	102.910	11.400	2
3	56.980	26.510	3
10	61.150	10.720	3
30	108.270	8.270	3
100	51.360	8.940	3
300	58.960	4.960	3
1000	21.160	9.250	3

Table 7: Mean of net proliferation with control set to 100%, GH3 cells, 48 hours treated with NaHS.

Table 7 shows the results of GH3 cells treated with NaHS for 48 hours. NaHS decreased dose dependently the net proliferation.

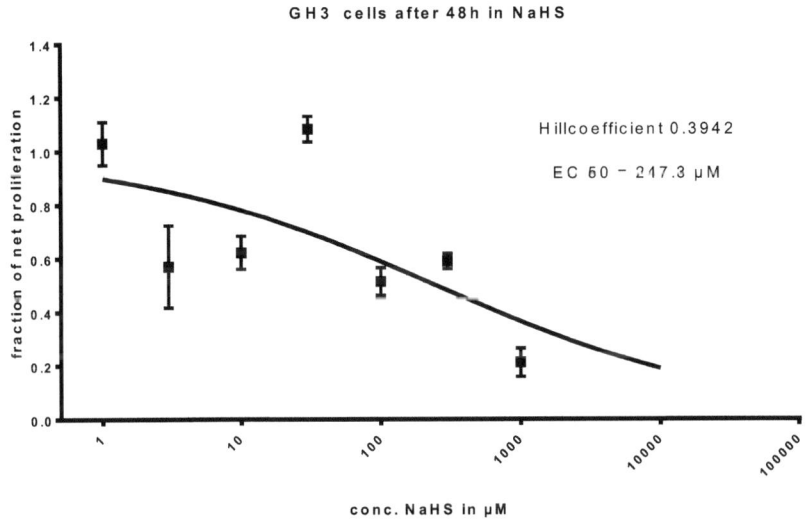

Figure 15: Dose response curve, GH3 cells treated for 48 hours with NaHS at various concentrations.

The data set from table 7 was used to create figure 15. An EC 50 value of 247.3 µM NaHS indicates that it cell reduces the net proliferation rate in GH3 cells by 50 %, when incubated with NaHS for 48 hours. A Hill coefficient of 0.3942 indicates no cooperative interaction between H_2S and its effect on GH3 cells. Concentrations of 1 µM and 30 µM NaHS seem to enhance cell proliferation, but the fact that 3 µM NaHS and 10 µM NaHS seem to be antiproliferative questions this conclusion. It rather may be that due to their lower attachment properties in the culture dish GH3 cells may have detached and were lost during the handling in the course of these experiments which may result in these variations at lower NaHS concentrations and the rather huge variance at 3 µM NaHS. Clearly more experiments are needed to improve these experimental data in question.

GH3 cells 24 hours incubation with NaHS and 24 hours recovery in NaHS free conditions:

Figures 16 and 17 show the cell counting data from GH3 cells after 24 hours with and without 1000 µM NaHS followed by a 24 hours recovery phase in NaHS free control medium. Both graphs are from the same experiment. In figure 16 one can see a very small peak around 33 µm. These big particles are most likely two or more cells stuck together.

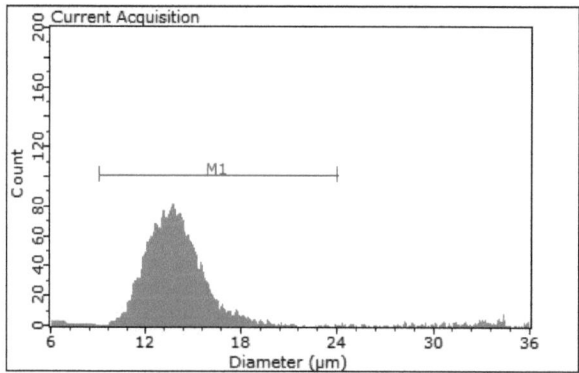

Figure 16: Cell counting data for GH3 cells 48 hours incubated in control medium, x axis represents cell diameter in µm, the y axis represents the number of cells counted. M1 indicates the data area used for calculating the total cell number.

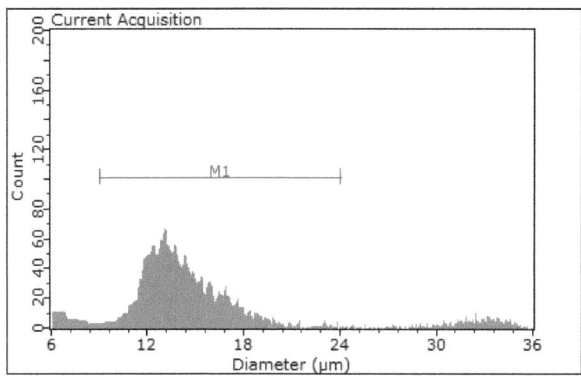

Figure 17: Cell counting data for GH3 cells incubated in 1000 µM NaHS for 24 hours followed by 24 hours recovery period in NaHS free control medium, x axis represents cell diameter in µm, the y axis represents the number of cells counted. M1 indicates the data area used for calculating the total cell number.

NaHS (µM)	mean net proliferation (%)	SD	N
0	100	-	3
30	76.15	19.60	3
1000	41.92	10.99	3

Table 8: Mean of net proliferation with control set to 100%, GH3 cells, 24 hours treated with NaHS and 24 hours recovery phase with NaHS free control medium.

In table 8 the results of the experiment is listed. GH3 cells were treated with NaHS for 24 hours with a 24 hour recovery phase in NaHS free control medium. The arithmetic mean of the net proliferation is given in percent. Due to limited time just two different concentrations were tested. These preliminary data suggest that NaHS reduces cell proliferation in GH3 cells like in C6 cells in a dose dependent manner.

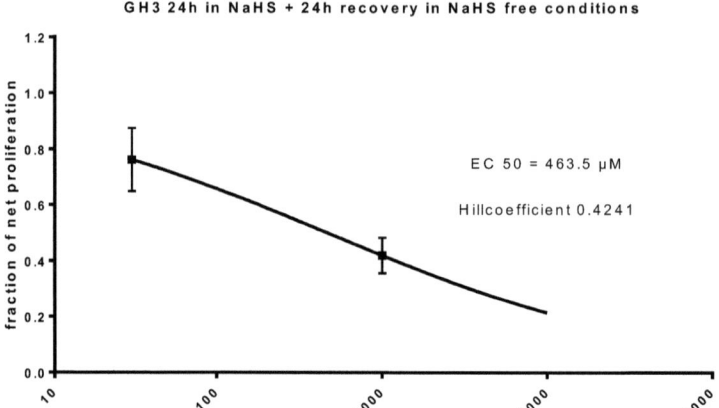

Figure 18: Dose response curve, GH3 cells treated for 24 hours with NaHS followed by a 24 hours recovery phase in NaHS free control media.

The data from table 8 were used to create figure 18. An EC 50 value of 463.5 μM NaHS indicates that GH3 cells reduce their net proliferation rate by 50 %, when incubated with NaHS for 48 hours. The Hill coefficient of 0.4241 indicates no cooperative interaction between the H_2S and an enzyme of the GH3 cells. Due to limited time just two data points could be obtained. As a consequence the curve, the EC 50 and the Hill coefficient in figure 18 have to be considered as very preliminary and approximate results.

Cell diameter and cell volume

Apart from the cell count, the Scepter technique also calculates cell diameter and cell volume.

C6 cells after 24 hours of incubation with NaHS followed by a 24 h hour recovery:

NaHS concentration (µM)	mean diameter control (µm) ± SD	mean diameter NaHS (µm) ± SD	N
3	13.81 ± 0.21	13.82 ± 0.18	3
10	13.75 ± 0.32	13.75 ± 0.13	3
30	13.69 ± 0.67	13.67 ± 0.59	3
100	14.19 ± 0.35	14.15 ± 0.28	4
300	13.68 ± 0.11	14.02 ± 0.08	3
1000	14.11 ± 0.13	14.53 ± 0.25	3
3000	13.73 ± 0.27	14.60 ± 0.29	4

Table 9: Diameter of C6 cells, treated for 24 hours with NaHS followed by a 24 hours recovery period in NaHS free control medium (SD= standard deviation).

The data for the cell diameter of C6 cells after 24 hours NaHS treatment followed by a 24 hour recovery period in NaHS free control medium (see table 9) reveal no significant difference between treated and control cells as statically tested with the paired samples t-test and the Wilcoxon signed ranks test. These tests have been used for all following statistical calculations.

NaHS concentration (µM)	mean volume control (pL) ± SD	mean volume NaHS (pL) ± SD	N
3	1.38 ± 0.05	1.39 ± 0.06	3
10	1.36 ± 0.09	1.36 ± 0.04	3
30	1.35 ± 0.20	1.34 ± 0.17	3
100	1.50 ± 0.15	1.48 ± 0.11	4
300	1.34 ± 0.08	1.44 ± 0.10	3
1000	1.47 ± 0.07	1.60 ± 0.12	3
3000	1.35 ± 0.21	1.63 ± 0.22	4

Table 10: Cell volume of C6 cells after incubation of cells for 24 hours in NaHS containing medium followed by a 24 hours recovery in NaHS free control medium (SD= standard deviation).

The data for the cell volume of C6 cells after incubation for 24 hours at various NaHS concentrations followed by 24 hours recovery with NaHS free control medium (see table 10) reveal no significant difference between treated and control cells.

C6 cells after 48 hours incubation with NaHS:

NaHS concentration (µM)	mean diameter control (µm) ± SD	Mean diameter NaHS (µm) ± SD	N
3	14.12 ± 0.20	14.10 ± 0.03	2
10	13.83 ± 0.28	13.96 ± 0.45	3
30	14.30 ± 0.31	14.18 ± 0.35	2
100	14.26 ± 0.40	14.09 ± 0.44	3
300	14.22 ± 0.22	14.33 ± 0.33	3
1000	14.08 ± 0.17	14.60 ± 0.05	4
3000	14.25 ± 0.35	16.16 ± 1.74	2

Table 11: Diameter of C6 cells after inbuation for 48 hours with various concentrations of NaHS (SD= standard deviation).

The data for the cell diameter of C6 cells after 48 hours NaHS treatment (see table 11) reveal no significant difference between treated and control cells.

NaHS concentration (µM)	mean volume control (pL) ± SD	mean volume NaHS (pL) ± SD	N
3	1.48 ± 0.05	1.47 ± 0.06	2
10	1.39 ± 0.08	1.42 ± 0.14	3
30	1.53 ± 0.07	1.49 ± 0.04	2
100	1.53 ± 0.09	1.46 ± 0.11	3
300	1.50 ± 0.12	1.54 ± 0.10	3
1000	1.46 ± 0.13	1.63 ± 0.15	4
3000	1.59 ± 0.03	2.75 ± 0.23	2

Table 12: Cell volume of C6 cells after incubation with various NaHS concentrations for 48 hours (SD= standard deviation).

The data for the cell volume of C6 cells after 48 hours NaHS treatment (see table 12) reveal no significant difference between treated and control cells.

GH3 cells after 48 hours incubation with various concentrations of NaHS:

NaHS concentration (µM)	mean diameter control (µm) ± SD	mean diameter NaHS (µm) ± SD	N	sig. dif.
1	13.95 ± 0.33	14.46 ± 0.37	2	Yes
3	14.25 ± 0.47	14.39 ± 0.70	5	No
10	14.64 ± 0.46	14.82 ± 0.60	3	No
30	14.66 ± 0.93	14.76 ± 0.88	3	No
100	14.13 ± 0.52	14.37 ± 0.38	4	No
300	14.41 ± 0.40	14.84 ± 0.69	3	No
1000	14.36 ± 0.33	15.23 ± 0.50	3	Yes

Table 13: Diameter of GH3 cells after an 48 hour period with various concentrations of NaHS (sig. dif.: significant difference between control and NaHS, SD= standard deviation).

The data for the cell diameter of GH3 cells treated with NaHS for 48 hours (see table 13) reveals that the diameter of the control cells is always a little bit smaller than the diameter of the cells treated with NaHS. A significant difference between control cells and treated cells was found for 1 µM and 1000 µM NaHS.

NaHS concentration (µM)	mean volume control (pL) ± SD	mean volume NaHS (pL) ± SD	N	sig. dif.
1	1.42 ± 0.17	1.59 ± 0.15	2	Yes
3	1.52 ± 0.23	1.56 ± 0.22	5	No
10	1.65 ± 0.15	1.71 ± 0.21	3	No
30	1.67 ± 0.34	1.70 ± 0.32	3	No
100	1.48 ± 0.16	1.56 ± 0.12	4	No
300	1.57 ± 0.13	1.72 ± 0.25	3	No
1000	1.55 ± 0.11	1.85 ± 0.18	3	Yes

Table 14: Cell volume of GH3 cells, 48 hours NaHS treatment (sig. dif.: significant difference between control and NaHS, SD= standard deviation).

The data for the cell volume of GH3 cells treated with NaHS for 48 hours (see table 14) reveals that the volume of the control cells is always a little bit smaller than the volume of the cells treated with NaHS. A significant difference between control cells and treated cells could be uncovered for 1 µM and 1000 µM NaHS.

GH3 net proliferation - 48 hours incubation with polyamines

First putrescine, spermine and spermidine were added in various concentrations, to determine their individual effect on the cell proliferation of GH3 cells.

putrescine (µM)	mean net proliferation (%)	SD	N
0	100	-	3
1000	98.32	18.715	3

Table 15: Mean of net proliferation with control set to 100%, GH3 cells were incubated with putrescine for 48 hours in serum free medium (Optimem).

In table 15 the net proliferation for GH3 cells when incubated with 1000 µM putrescine is shown. It was found that even the addition of 1000 µM putrescine did not alter the proliferation rate significantly compared to control. As a consequence of this indifferent result no further experiments with different concentrations have been performed.

spermidine (µM)	mean net proliferation (%)	SD	N
0	100	-	5
3	89.90	19.42	5
30	-57.69	60.87	5

Table 16: Mean of net proliferation with control set to 100%, GH3 cells and were incubated with 3 or 30 µM spermidine for 48 hours in serum free medium (Optimem).

Spermidine shows a strong reduction in net proliferation at a concentration of 30 µM (see table 16). However, the negative net proliferation indicates that the final cell count after 48 hours incubation with 30 µM spermidine was less than the amount of initially seeded cells. The high standard deviation (SD) of 60,87 indicates a great fluctuation in this experiment, despite a N of 5. This result was found to be statistically significant when tested with a Wilcoxon signed ranks test. At 3 µM spermidine the effect was less pronounced but also statistically significant.

spermidine (µM)	mean cell diameter control (µm) ± SD	mean cell diameter spermidine (µm) ± SD	N
3	13.50 ± 0.49	13.63 ± 0.36	5
30	13.22 ± 0.49	14.32 ± 1.65	5

Table 17: cell diameter of GH3 cells after 48 hour treatment with spermidine.

spermidine (µM)	mean cell volume control (pL) ± SD	mean cell volume spermidine (pL) ± SD	N
3	1.30 ± 0.15	1.33 ± 0.10	5
30	1.22 ± 0.14	1.66 ± 0.57	5

Table 18: cell volume of GH3 cells after 48 hour treatment with spermidine.

Cell diameter (table 17) and cell volume (table 18) are statistically significant enhanced (tested with the Wilcoxon signed ranks test and paired samples t-test) in GH3 cells after treatment with 3 and 30 µM spermidine for 48 hours. Together with the reduction in cell proliferation (table 25), the reduction in cell diameter and volume after 48 hour incubation with 3 or 30 µM spermidine may indicate that apoptosis is imminent in these cells.

spermine (µM)	Mean Net proliferation (%)	SD	N
0	100	-	-
30	-36.63	13.045	2

Table 19: Mean of net proliferation with control set to 100%, GH3 cells and spermine for 48 hours.

In table 19 the addition of 30 µM spermine to the GH3 cells for 48 hours is depicted. The negative net proliferation indicates that the final cell count after 48 hours incubation with 30 µM spermine was less than the amount of initially seeded cells. No more experiments were performed because of the great similarity between spermidine and spermine and because of the limited time.

spermine (µM)	mean cell diameter control (µm) ± SD	mean cell diameter spermine (µm) ± SD	N
30	13.93 ± 0.55	14.28 ± 0.80	2

Table 20: cell diameter of GH3 cells after 48 hour treatment with spermine.

spermine (µM)	mean cell volume control (pL) ± SD	mean cell volume spermine (pL) ± SD	N
30	1.42 ± 0.17	1.54 ± 0.26	2

Table 21: cell volume of GH3 cells after 48 hour treatment with spermine.

Cell diameter (table 20) and cell volume (table 21) were non significantly enhanced (tested with Wilcoxon signed ranks test) when treated with spermine for 48 hours.

A combined incubation of GH3 cells with polyamines (3 µM spermidine) and H_2S (30 µM NaHS) was found to most likely reduce cell proliferation and viability further. Due to limited time no representative data could be obtained for this experimental setting.

Calcium Imaging

The Ca^{2+} imaging data were offline transferred to an Excel file and then plotted in Excel for each experiment individually - for example see figure 18. Each graph displays first the base line (about 60 seconds), followed by the application of 100 µM NaHS, for about 15 seconds, followed by about 30 seconds of control solution (= control 2). Finally a physiological solution containing 15 mM KCl was applied for 15 seconds followed by control solution for washout. The KCl solution should provoke a "survival" Ca^{2+} signal expected to be seen only in living cells. Only cells which showed the KCl response at the end of the experiment were considered to be alive and were further analyzed.

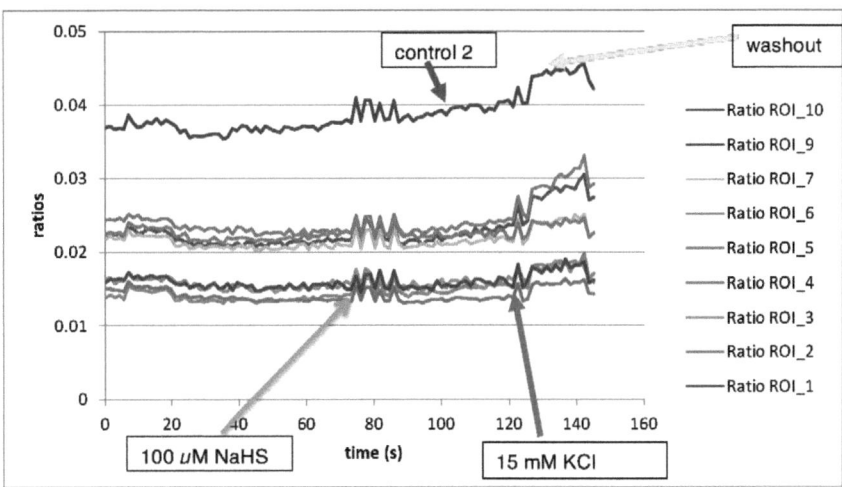

Figure 19: Original traces of a Ca^{2+} imaging experiment in GH3 cells. 100μM NaHS increase intracellular calcium concentrations and evoke transient Ca^{2+} oscillations. Arrows indicate the time points when NaHS or KCl was added to the perfusate or when control solutions were applied for washout purposes. Each trace represents the ratio of a distinctive ROI (as generated from figure 4) (graph created with Microsoft Excel).

The traces in figure 19 are shown to illustrate the reaction of the GH3 cells to NaHS. Shortly after the application of 100 μM NaHS an ozillation of the calcium signal was observed in 10 cells out of 349 cells which reacted positive when NaHS was applicated. All others cells which reacted positive to NaHS show a different response (example: figure 20). The response to 15 mM KCl was the same in all experiments resulting in an increase of intracellular calcium.

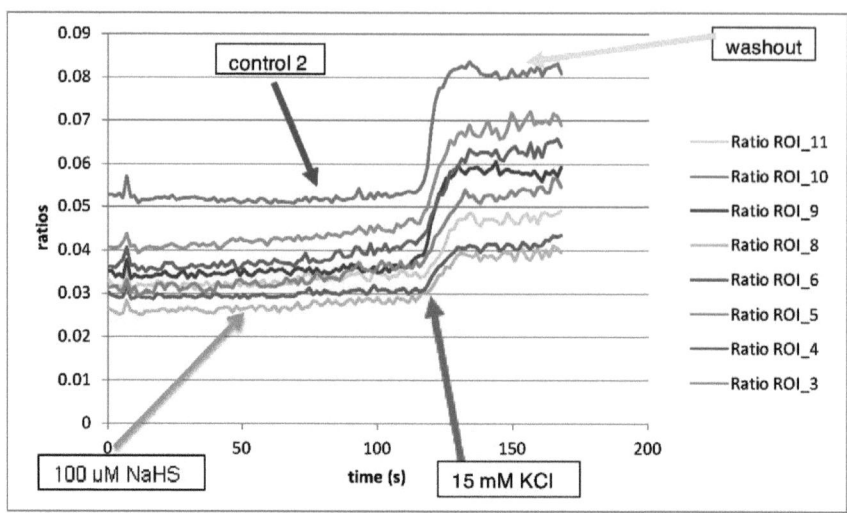

Figure 20: Original traces of a Ca²⁺ imaging experiment in GH3 cells. 100μM NaHS increases intracellular calcium concentrations. Arrows indicate the time points when NaHS or KCl was added to the perfusate or when control solutions were applied for washout purposes. Each trace represents the ratio of a distinctive ROI (as generated from figure 3) (graph created with Microsoft Excel). In these cells no Ca²⁺ oscillations were detected.

A total of 349 cells were analyzed to calculate the effect of H_2S on intracellular Ca^{2+} levels:

Descriptive Statistics

	N	Minimum	Maximum	Mean	Std. Deviation
Control	349	0.01549	0.17552	0.0305855	0.01414079
100 μM NaHS	349	0.01520	0.15992	0.0318734	0.01407674
control 2	349	0.01533	0.15385	0.0329544	0.01418981
15 mM KCl	349	0.01595	0.17173	0.0378660	0.01639736
Washout	349	0.01750	0.18890	0.0450585	0.01878023
Valid N (listwise)	349				

Table 22: Descriptive statistics of calcium imaging data, numbers represent the fluorescent ratios measured and calculated by the acquisition software (Version: 2.4.0.15) from FEI, Munich (the table was created with SPSS software).

Fluorescence ratios were calculated for control, 100 μM NaHS, the second control solution application (control 2) 15 mM KCl, and the washout response. 37 cells showed a negative response, 156 a positive and 156 no response to NaHS treatment. The response was considered positive or negative, if the difference from control to NaHS was ≥0,001. To assure a normal distribution of the data, the log(10) of every data point was calculated. The negative responses were excluded for the calculation, on the one hand because otherwise no normal distribution of the data is guaranteed, and on the other hand because a negative Ca^{2+} response may be an indicator for a damaged or not normal functioning cell. Additionally three outliers were eliminated, after performing a Grubbs test for outliers (cell numbers 71, 138 and 269). The fluorescence ratio data from the remaining 309 cells (see table 22) were logarithmised (see table 23) and tested for normality distribution (see table 24). Kolmogorov-Smirnov and Shapiro-Wilk tests show a normal distribution of the data. Hence a paired samples t-test was performed (see table 25).

Descriptive Statistics

	N	Minimum	Maximum	Mean	Std. Deviation
control	309	0.01549	0.05312	0.0284708	0.00724407
100 µM NaHS	309	0.01520	0.07694	0.0302614	0.00877710
control 2	309	0.01533	0.07240	0.0314675	0.00929849
15 mM KCl	309	0.01595	0.08452	0.0364240	0.01162005
washout	309	0.01750	0.09416	0.0436357	0.01436668
Valid N (listwise)	309				

Table 23: Descriptive statistics of calcium imaging data, numbers represent the fluorescent ratios measured (the table was created with SPSS software).

Descriptive Statistics

	N	Minimum	Maximum	Mean	Std. Deviation
control	309	-1.80992	-1.27476	-1.5588519	0.10658665
100 µM NaHS	309	-1.81809	-1.11383	-1.5359193	0.11979780
control 2	309	-1.81443	-1.14027	-1.5198587	0.12342482
15 mM KCl	309	-1.79737	-1.07305	-1.4595526	0.13469556
Washout	309	-1.75709	-1.02613	-1.3834421	0.14341767
Valid N (listwise)	309				

Table 24: Descriptive statistics of calcium imaging data; data were logarithmised (created with SPSS).

Tests of Normality

	Kolmogorov-Smirnov[a]			Shapiro-Wilk			Normality given (Yes/No)
	Statistic	df	Sig.	Statistic	df	Sig.	
control	0.040	309	0.200*	0.993	309	0.146	Yes
100 µM NaHS	0.040	309	0.200*	0.995	309	0.347	Yes
control 2	0.026	309	0.200*	0.995	309	0.485	Yes
15 mM KCl	0.034	309	0.200*	0.996	309	0.688	Yes
Washout	0.032	309	0.200*	0.994	309	0.281	Yes

*. This is a lower bound of the true significance.

a. Lilliefors Significance Correction

Table 25: Test of normality for Ca^{2+} imaging data from table 22 (Statistic=T-value (one tailed), df=degrees of freedom, Sig.=significance level (one tailed)) (created with SPSS).

Paired Samples Test

		Paired Differences					t	df	Sig. (2-tailed)	Significant (Yes/No)
		Mean	Std. Deviation	Std. Error Mean	99% Confidence Interval of the Difference					
					Lower	Upper				
Pair 1	control -100 µM NaHS	-0.02293253	0.03374915	0.00191992	-0.02790875	-0.01795631	-11.945	308	0.000	Yes
Pair 2	control – 15 mM KCl	-0.09929931	0.06971739	0.00396608	-0.10957894	-0.08901967	-25.037	308	0.000	Yes
Pair 3	control - control 2	-0.03899316	0.04281622	0.00243573	-0.04530629	-0.03268002	-16.009	308	0.000	Yes
Pair 4	control - Washout	-0.17540979	0.08381020	0.00476780	-0.18776738	-0.16305221	-36.791	308	0.000	Yes

Table 26: Paired t-test for Ca^{2+} imaging data (99% confidence interval), logarithmised fluorescence ratios from table 22 (df=degrees of freedom, t=T-value, Sig.=significance level (two tailed)) (table created with SPSS software).

To visualize the data distribution, I drew a box and whiskers blot (figure 21), using Graphpad software.

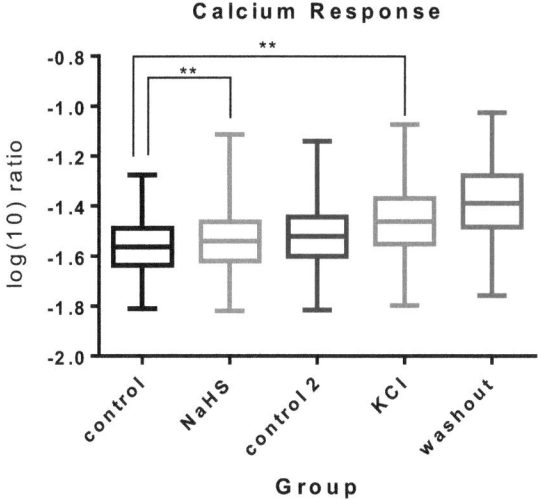

Figure 21: Box and whiskers blot of Ca^{2+} imaging data from table 10; x axis represents the different groups, y axis the logarithmized calcium imaging ratios (** indicates a highly significant difference between the marked groups) (created with Graphpad Prism software).

group	mean ratio	%	% increase after addition
control	0.028471	100	0
NaHS	0.030261	106.29	6.29
control 2	0.031468	110.53	10.53
KCl	0.036424	127.93	27.93
washout	0.043636	153.26	53.26

Table 27: Mean ratio of calcium imaging data, percental increase in relation to control is shown.

Table (27) shows the mean ratio of the 349 analyzed cells. I calculated from this the percental increase in relation to the control level (100 %). The addition of NaHS resulted in a 6.29 % enhancement of the Ca^{2+} ratio. The subsequent application of control 2 shows an even greater response (10.53 %), which may be interpreted as a delayed or priming response to NaHS. Further experiments should investigate if a longer application of control-2 leads again to control values. KCl and washout show the greatest enhancement in ratio addition. Since KCl depolarizes the membrane potential an increase of the intracellular Ca^{2+} is expected, caused by the opening of Ca^{2+} channels. However, the subsequent washout again resulted in an even further increase of the fluorescence ratio which is difficult to explain but may be a sign that the KCl treatment irreversibly damaged the cells.

Discussion

Hydrogen sulfide

Searching the literature databank Pubmed (http://www.ncbi.nlm.nih.gov, accessed 22.July 2014), one can find publications with proliferation enhancing (Chao et al. 2013; Fan *et al.* 2013; Hellmich et al. 2014 Hong et al. 2014; Pan et al. 2014; Vadivel et al. 2014) as well as with proliferation decreasing (Lv et al. 2014) effects caused by H_2S. These contradictory results may be the consequence of H_2S toxicity on the one side and its essentiality for life on the other side. To verify the effect of H_2S on the cell proliferation, I applied NaHS in various concentrations to C6 and GH3 cells. The cells were either incubated with NaHS for 48 or for 24 hours followed a by a 24 hour recovery period. Afterwards the cells were counted and the net proliferation rate was determined. NaHS reduced the proliferation rate in a concentration dependent manner in both cell lines. The data further show that an incubation for 24 hours with NaHS followed a 24 hour recovery period is by far less toxic than a continuous incubation for 48 hours with NaHS.

GH3 cells incubated for 48 hours with NaHS have a much smaller EC50 value (247.3 µM) compared to C6 cells (890.8 µM) when incubated for 48 hours with NaHS, This suggests that GH3 cells are more susceptible to H_2S than C6 cells. On the other hand the experiments with GH3 cells were more difficult to conduct, due to the fact that GH3 cells detach easily from the cell culture dish. As seen in figure 6 (especially at 30 µM NaHS), the variance is relatively high, compared to the C6 experiments.

This can be interpreted as a result of detached cells and thus a possible loss of cells during the experiment.

Hill values of 1,42 (when incubated for 48 hours with NaHS) and 0,58 (when incubated for 24 hours with NaHS and a 24 hour recovery phase) for C6 and 0,39 (when incubated for 48 hours with NaHS) and 0,42 (when incubated for 24 hours with NaHS and a 24 hour recovery phase) for GH3 indicate in both cells lines no cooperation with H_2S, except when C6 cells are incubated for 48 hours with NaHS. I interpret this particular result as an indicator for a cooperative reaction, which means that the affinity for an additional molecule is greater for each prior bond molecule. On the other hand a Hill value of 1.4 indicates a rather nonspecific action of H_2S, since this value is pretty much in the middle between 1 and 2. All other Hill values are less than 1 and therefore represent no cooperation between a possible target site and H_2S indicating a rather unspecific action of H_2S. In this context it would be interesting to study how H_2S interacts with the cells in detail.

Proliferation remains almost at control level at low H_2S concentrations (3 µM till 10 µM). However, at 30 µM NaHS C6 cells which were incubated for 48 hours with NaHS show an increasing net proliferation which was non-significant. This could be due to the small sample size, however.

To obtain the actual H_2S concentration, one has to divide the NaHS concentration by 3 (Reiffenstein et al., 1992) or probably by about 8 if the salinity of the solution and evaporation is taken into account (Sitdikova, Weiger and Hermann 2014, data from our lab in preparation). Salinity of the solution and evaporation of H_2S depletes the actual H_2S concentration additionally and has to be taken into account. The calculations from the unpublished data of our lab show that just 11-13% of the applied NaHS is available as H_2S.

Huang et al. (2010) found, that a concentration of 15 µM NaHS stimulates the proliferation of interstitial cells of Cajal (gastrointestinal cells) and Liu et al. (2014) also concluded, that NaHS has a positive effect on proliferation of neuronal stem cells (NSC) at low concentrations, namely 0,1 to 50 µM NaHS. Cai et al. (2010) found that H_2S promotes proliferation of human colon cancer cells with best results obtained at a concentration of 200 µM NaHS. Molecular methods uncovered, that the phosphorylation of Akt and ERK was enhanced in cells treated with NaHS. Blocking Akt and ERK phosphorylation caused no proliferation enhancement (Cai et al. 2010).

Chattopadhyay et al., (2013) used H_2S releasing aspirin (HS-ASA) as a donor. They found that this donor decreased proliferation of acute lymphoblastic leukemia. Notably the cells were locked in G0/G1 cell cycle checkpoint, which subsequently induced apoptosis and decreased proliferation.

In contrast to the above cited literature decreased cell proliferation was found by Fan et al. (2013). They added 500 µM NaHS to HSC-T6 liver rat cells. The authors concluded that the cell cycle arrest in G1 was mediated by an increased phospho-Akt expression and a decreased expression of phospho-p38. The Akt/GSK3β (glycogen synthase kinase 3 β) signaling pathway (which is crucial in cell cycle and apoptosis) also plays a significant role in the results by Han et al. (2013). They found lymphocytes to increase proliferation at "low" concentrations (<1 mM) of NaHS and to decrease proliferation at higher concentrations (>2 mM). Even though the concentrations used in these studies are higher than in my and in most other findings, they support the notion that NaHS is on the one hand inhibiting the proliferation at high concentrations and on the other hand may enhance proliferation at low concentrations. Due to the fact that my data does not show significant differences for these concentrations more experiments are needed to test this hypothesis.

C6 and GH3 cells both show decreasing net proliferation with increasing NaHS concentration. Application of 3000 µM NaHS had a huge impact on cell proliferation. However, one should keep in mind that this is an unphysiological high concentration, where cyctochrome C oxidase is most likely inhibited (Hermann et al. 2010). C6 cells incubated for 24 hours with 3000 µM NaHS followed by a 24 hours release have a decreased proliferation rate of about 60 %, while the cells incubated 48 hours with NaHSs have a net proliferation rate around zero. This suggests that the C6 cells can recoup from the toxic NaHS treatment when allowed to recover in a control physiological solution for 24 hours.

The data from the experiments with GH3 cells are far more scattered than the data from C6 cells. As mentioned earlier GH3 cells are more difficult to handle because they detach easily from the flask which is possibly resulting in a larger variance of the data.

The effect of H_2S on cell proliferation in my investigations is comparable other results reported in the literature (Cai *et al.*, 2010; Chattopadhyay *et al.*, 2013; Fan *et al.*, 2013; Han *et al.*, 2013; Huang *et al.*, 2010; Li *et al.*, 2012; Liu *et al.*, 2014; Perry *et al.*, 2011; Wu *et al.*, 2012; Yang and Wang, 2007; Zhao *et al.*, 2013) showing that H_2S causes a decrease in cell proliferation in a dose dependent manner.

Interestingly GH3 cells treated with 1 µM and 1000 µM NaHS for 48 hours have a significantly increased cell diameter as well as an increased volume. Even though all other concentrations did not show a significant difference, this could be a hint for beginning apoptosis, which is characterized through an increase in cell diameter and volume (Löffler et al. 2008; Alberts 2002). Due to limited data in the experiment with GH3 cells under 24 hours treatment with NaHS followed by 24 hours recovery, no significant results could be obtained for cell diameter and volume. All experiments with C6 cells on the other hand did not show a significant increase in cell diameter and volume when treated with NaHS.

C6 and GH3 cells both originate from rat brain but the response to H_2S is different. As mentioned above different cell types often have different reactions to H_2S. In addition proliferation enhancing and decreasing effects may depend on the concentrations applied.

Polyamines

Polyamines play a crucial role in the cell cycle. Their concentrations vary during the cell cycle and have three peaks (Alm and Oredsson, 2009). The first peak is at the switch from G0 to G1 phase, the second at the change from G1 to S phase and the third in the G2 phase. This indicates a strong link to cell cycle control, because all three peaks are at crucial checkpoints in the cell cycle, where the cell has to decide to move on in the cell cycle or stop the reproduction (Alm and Oredsson, 2009).

My hypothesis that polyamines are necessary for the cell cycle and may enhance cell proliferation, based on the findings of Alm and Oredsson (2009), was however not supported by my data. All experiments in my study conducted with polyamines except for putrescine resulted in a decrease in viability, and proliferation rates. Spermidine also increased the cell diameter and the cell volume of GH3 cells after 48 hours incubation with 3 or 30 µM of spermidine. This may be is also an indicator for apoptosis (Löffler et al. 2008; Alberts 2002). Interestingly other authors (as reviewed in Moschou and Roubelakis-Angelakis, 2014; Weissel *et al.* 2014; Casero and Marton, 2007) found that polyamines (at concentrations ranging from 25 µM to 5 mM) and their metabolic derivates may induce programmed cell death which fits my results. But like with NaHS the effect of polyamines depends on the concentration used.

Calcium imaging

Ca^{2+} is one of the most important signaling molecule in living cells and also plays a crucial role in cell proliferation, apoptosis and survival of the cells (Borowiec et al. 2013, Capiod 2013). Calcium imaging can give an important insight in cellular mechanisms and the underlying pathways (Parekh and Penner, 1997; Parys *et al.*, 2014). NaHS was applied during Ca^{2+} imaging experiments to uncover the effect of H_2S on intracellular Ca^{2+}. Cells showing a negative response were excluded from calculations. They were considered to be dead or dying cells and therefore they were not included in the data analysis. Interestingly 10 out of 349 analyzed cells showed Ca^{2+} oscillations in the response to 100 µM NaHS. This could be either an artifact of this specific experiment or it could be evidence for adaptation of the GH3 cells to H_2S. Searching the literature databank Pubmed (http://www.ncbi.nlm.nih.gov accessed 8. August 2014) current literature concerning this matter could be found. Kaneko *et al.* (2006) found that L-cystein and NaHS reversibly reduced glucose-induced Ca^{2+} oscillation in mouse pancreatic cells. NaHS significantly enhanced the intracellular Ca^{2+} concentration, even though the difference is quite small, about 6 % when Fura-2 ratios were compared. The difference between control and KCl is much higher (27 %), indicating viability of the cells.

A similar experiment was reported by Pupo *et al.* (2011). They applied NaHS to different endothelial cell types at concentrations from 1 nM to 3 mM. The most prominent effects were observed with concentrations of 1 µM in BTECs (rat breast lobular-infiltrating carcinoma cells) and 3 mM NaHS in rat aortic endothelial cells.

Avanzato *et al.* (2014) applied NaHS (10 µM) and Nifedipine (10 µM), a Ca^{2+}-channel blocker, to cardiomyocytes. They found that NaHS (10 µM) or Nifedipine (10 µM) decreased free intracellular calcium concentrations. This suggests that L-type voltage-operated calcium channels are blocked by H_2S. In the simultaneous presence of Ni^{2+} (100 µM) and Nifedipine H_2S had no effect any more on internal Ca^{2+} concentrations indicating that next to L-Type also T-type calcium channels are involved in the action of H_2S. Even though this would explain the negative responses in my experiments, one has to consider, that Avanzato et al. used cardiomyocytes, which may respond differently to the cells I used. Pretreatment of cardiomyocytes with NaHS (10 µM) prevented H_2O_2 induced cell death in their system indicating a beneficial aspect of H_2S under oxidative stress situations.

Summary

My experiments indicate that H_2S acts on cell proliferation in a concentration dependent manner. Low NaHS concentrations (<30 µM which is resulting in a effective H_2S concentration of approximately 3,75 µM) may enhance proliferation and high concentrations (>300 µM resulting in an effective H_2S concentration of approximately 37,5 µM) reduces the proliferation rate. Two different neuronal cell types (C6 and GH3) were similar affected by H_2S. In additionGH3 cells showed a significant increase in cell diameter and volume when incubated with 1 µM and 1000 µM NaHS for 48 hours, suggesting that NaHS induces apoptosis in GH3 cells. Further experiments are needed to test this hypothesis. The results are in agreement with findings and conclusions of other groups: (Cai et al., 2010; Chattopadhyay et al., 2013; Fan et al., 2013; Han et al., 2013; Huang et al., 2010; Li et al., 2012; Liu et al., 2014; Perry et al., 2011; Wu et al., 2012; Yang and Wang, 2007; Zhao et al., 2013).

Spermidine and spermine reduced the cell viability of GH3 cells, when incubated for 48 hours. Putrescine seems to have no such effect. Spermidine also increased the cell diameter and volume of GH3 cells, suggesting that spermidine may induce apoptosis in GH3 cells. Again, these findings are in agreement with results of other groups (as reviewed in Moschou and Roubelakis-Angelakis 2014; Weissel et al. 2013; and Casero and Marton, 2007)

Ca^{2+} imaging confirmed the importance of H_2S for modulating internal Ca^{2+} concentrations. Intracellular Ca^{2+} was significantly increased after the application of 100 µM NaHS. Similar findings were reported by other groups (Liang et al., 2012; Parys et al., 2014; Pupo et al., 2011).

Future Outlook

My data underline the importance of H_2S modulating cell proliferation. More research will be needed to fully understand the underlying mechanisms and pathways. A new and interesting H_2S dye may help to visualize the distribution of H_2S inside living cells (Chen *et al.*, 2013a). Combined with other dyes this could help to unravel crucial interactions of H_2S with cellular molecules. Another important experiment would be to quantify the intracellular Ca^{2+} changes Molecular tools such as PCR, Western blotting or electrophysiological methods should also be used to further study the effects of H_2S. As other groups have shown, these tools are essential in identifying the molecular targets of H_2S and their modulation by H_2S. Conducting further experiments with H_2S and polyamines could help to further understand the connection between these crucial factors for cell proliferation.

Literature

Abe, K. and Kimura, H. (1996), "The possible role of hydrogen sulfide as an endogenous neuromodulator", *The Journal of Neuroscience*, pp. 1066–1071.

Alberts, B. (2002), *Molecular biology of the cell,* 4th ed, Garland Science, New York.

Ali, M.Y., Ping, C.Y., Mok, Y.Y. (2006), "Regulation of vascular nitric oxide in vitro and in vivo; a new role for endogenous hydrogen sulphide?", *Br. J.Pharmacol.*, Vol. 149, pp. 625–34.

Alm, K. and Odresson, S. (2009), "Cells and polyamines do it cyclically", *Essays in Biochemistry,* Vol. 46, pp. 63-76.

Avanzato, D., Merlino, A., Porrera, S., Wang, R., Munaron, L., Mancardi, D. (2014), "Role of calcium channels in the protective effect of hydrogen sulfide in rat cardiomyoblasts", *Cellular Physiology and Biochemistry*, Vol. 33 No. 4, pp. 1205-1214.

Borowiec, A.S., Bidaux, G., Pigat, N., Goffin, V., Bernichtein, S., Capiod, T. (2013), "Calcium channels, external calcium concentration and cell proliferation" , *European Journal of Pharmacology*, Vol. 739, pp. 19-25.

Bruintjes, J.J., Henning, R.H., Douwenga, W. and van der Zee, E A (2014), "Hippocampal cystathionine beta synthase in young and aged mice", *Neuroscience letters*, Vol. 563, pp. 135–139.

Capiod, T. (2013), "The need for calcium channels in cell proliferation" , *Recent Patents on Anti Cancer Drug Discovery*, Vol. 8, pp. 1-17.

Casero, Robert A.; Marton, Laurence J. (2007), "Targeting polyamine metabolism and function in cancer and other hyperproliferative diseases" , *Nature reviews. Drug discovery*, Vol. 6 No. 5, pp. 373-390.

Chao, .C, Coletta, C., Módis, K., Papapetropoulos, A., Szabo, C., Hellmich, M. (2013), "Cystathionine-β-synthase (CBS)-derived hydrogen sulfide (H2S) supports colorectal tumor growth and angiogenesis in vivo", *Nitric Oxide*, Vol. 31, pp. 37.

Chen, W.-L., Xie, B., Zhang, C., Xu, K.-L., Niu, Y.-Y., Tang, X.-Q., Zhang, P., Zou, W., Hu, B. and Tian, Y. (2013), "Antidepressant-like and anxiolytic-like effects of hydrogen sulfide in behavioral models of depression and anxiety", *Behavioural pharmacology*, Vol. 24 No. 7., pp. 590-597.

Chernikov, A.P. (1952), "Intoxications with hydrogen sulfide and their prevention", *Fel'dsher i akusherka*, Vol. 7, pp. 19–23.

Clark, R.A.F., McCoy, G.A., Folkvord, J.M., McPherson, J.M. (1998), "TGF-β1 stimulates cultured human fibroblasts to proliferate and produce tissue-like fibroplasia: A fibronectin matrix-dependent event" , *Journal of Cellular Physiology*, Vol. 170 No.1, pp. 69-80.

Elsey, D.J., Fowkes, R.C. and Baxter, G.F. (2010), "Regulation of cardiovascular cell function by hydrogen sulfide (H 2 S)", *Cell Biochemistry and Function*, Vol. 28 No. 2, pp. 95–106.

Fan, H. N; Wang, H. J, Ren, L., Ren, B., Dan, C. R. Y., Li, Y. F, Hou, L. Z, Deng, Y. (2013), „Decreased expression of p38 MAPK mediates protective effects of hydrogen sulfide on hepatic fibrosis", *European Review for Medical and Pharmacological Sciences*, Vol. 17, pp. 644-652.

Freidrich, A.W. (1946), "Hydrogen sulfide poisoning; report of two cases, one with fatal outcome, from associated mechanical asphyxia", *The American journal of pathology*, Vol. 22, pp. 147–155.

Giuliani, D., Ottani, A., Zaffe, D., Galantucci, M., Strinati, F., Lodi, R. and Guarini, S. (2013), "Hydrogen sulfide slows down progression of experimental Alzheimer's disease by targeting multiple pathophysiological mechanisms", *Neurobiology of Learning and Memory*, Vol. 104, pp. 82–91.

Grynkiewicz, G., Poenie, M. and Tsien, R.Y. (1985), "A New Generation of Ca2+ Indicators with Greatly Improved Fluorescence Properties", *Journal of Biological Chemistry,* Vol. 260, pp. 3440–3450.

Gutiérrez-Martín, Y., Martín-Romero, F.J., Henao, F., Gutiérrez-Merino, C. (2005), "Alteration of cytosolic free calcium homeostasis by SIN-1: High sensitivity of L-type Ca2+channels to extracellular oxidative ⁄ nitrosative stress in cerebellar granule cells", *J. Neurochem*, Vol. 92, pp. 973–89.

Han, Y., Zeng, F., Tan, G., Yang, C., Tang, H., Luo, Y., Feng, J., Xiong, H. and Guo, Q. (2013), "Hydrogen Sulfide Inhibits Abnormal Proliferation of Lymphocytes via AKT/GSK3β Signal Pathway in Systemic Lupus Erythematosus Patients", *Cellular Physiology and Biochemistry*, Vol. 31 No. 6, pp. 795–804.

He, X.-L., Yan, N., Zhang, H., Qi, Y.-W., Zhu, L.-J., Liu, M.-J. and Yan, Y. (2014), "Hydrogen sulfide improves spatial memory impairment and decreases production of Aβ in APP/PS1 transgenic mice", *Neurochemistry international*, Vol. 67, pp. 1–8.

Hellmich, M.R., Coletta, C., Chao, C., Szabo, C. (2014), "The Therapeutic Potential of Cystathionine β-Synthetase/Hydrogen Sulfide Inhibition in Cancer" , *Antioxid Redox Signal*, ahead of print.

Hermann, A., Sitdikova, G.F. and Weiger, T.M. (Eds.) (2012), *Gasotransmitters: Physiology and Pathophysiology*, Springer Berlin Heidelberg, Berlin, Heidelberg.

Hermann, A., Sitdikova, G.F. and Weiger,T.M. (2011), „Giftige Geister", *Gehirn und Geist,* Vol. 5, pp. 28-35.

Hermann, A., Sitdikova, G.F. and Weiger,T.M. (2010), „Gasotransmitter, Gase als zelluläre Signalstoffe", *Biologie in unserer Zeit,* Vol. 40 No. 3, pp. 185-193.

Hong, M., Tang, X. and He, K. (2014), "Effect of hydrogen sulfide on human colon cancer SW480 cell proliferation and migration in vitro", *Nan Fang Yi Ke Da Xue Xue Bao,* Vol. 34 No. 5, pp. 699-703.

Hu, L.-F., Wong, P.T.-H., Moore, P.K. and Bian, J.-S. (2007), "Hydrogen sulfide attenuates lipopolysaccharide-induced inflammation by inhibition of p38 mitogen-activated protein kinase in microglia", *Journal of Neurochemistry,* Vol. 100 No. 4, pp. 1121–1128.

Igarashi, K. and Kashiwagi, K. (2010), "Modulation of cellular function by polyamines", *The International Journal of Biochemistry & Cell Biology,* Vol. 42 No. 1, pp. 39–51.

Igarashi, K. and Kashiwagi, K. (2010), "Characteristics of cellular polyamine transport in prokaryotes and eukaryotes", *Plant physiology and biochemistry PPB / Société française de physiologie végétale,* Vol. 48 No. 7, pp. 506–512.

Kaneko, Y., Kimura, Y., Kimura, H., Niki, I. (2006), " l-Cysteine Inhibits Insulin Release From the Pancreatic β-Cell Possible Involvement of Metabolic Production of Hydrogen Sulfide, a Novel Gasotransmitter", *Diabetes,* Vol. 55 No. 5, pp. 1391-1397.

Kimura, H. (2010), "Hydrogen Sulfide: From Brain to Gut", *Antioxidants & Redox Signaling,* Vol. 12 No. 9., pp. 1111-1123.

Kimura, H. (2012), "Physiological and Pathophysiological Functions of Hydrogen Sulfide", in Hermann, A., Sitdikova, G.F. and Weiger, T.M. (Eds.), *Gasotransmitters: Physiology and Pathophysiology,* Springer Berlin Heidelberg, Berlin, Heidelberg, pp. 71–98.

Kimura, Y., Kimura, H. (2004), "Hydrogen sulfide protects neurons from oxidative stress", *FASEB Journal,* Vol. 18, pp. 1165–7.

Lee, Z.W., Zhou, J., Chen, C.-S., Zhao, Y., Tan, C.-H., Li, L., Moore, P.K. and Deng, L.-W. (2011), "The slow-releasing hydrogen sulfide donor, GYY4137, exhibits novel anti-cancer effects in vitro and in vivo", *PloS one,* Vol. 6 No. 6, pp. e21077.

Leffler, C.W., Parfenova, H., Basuroy, S., Jagger, J.H., Umstot, E.S., Fedinec, A.L. (2011), "Hydrogen sulfide and cerebral microvascular tone in newborn pigs", *American Journal of Physiology Heart and Circulatory Physiology,* Vol. 300 No. 3, pp. 440-447.

Li, M.-H., Tang, J.-P., Zhang, P., Li, X., Wang, C.-Y., Wei, H.-J., Yang, X.-F., Zou, W. and Tang, X.-Q. (2014), "Disturbance of endogenous hydrogen sulfide generation and endoplasmic reticulum stress in hippocampus are involved in homocysteine-induced defect in learning and memory of rats", *Behavioural brain research*, Vol. 262, pp. 35–41.

Li, Z.-y., Liu, S.-c., Xu, P.-j., Yang, Z. and Zhang, T. (2012), "[Hydrogen sulfide stimulates the development of rat glioblastoma]", *Zhonghua zhong liu za zhi [Chinese journal of oncology]*, Vol. 34 No. 4, pp. 254–258.

Liu, D., Wang, Z., Zhan, J., Zhang, Q., Wang, J., Zhang, Q., Xian, X., Luan, Q. and Hao, A. (2014), "Hydrogen sulfide promotes proliferation and neuronal differentiation of neural stem cells and protects hypoxia-induced decrease in hippocampal neurogenesis", *Pharmacology Biochemistry and Behavior*, Vol. 116, pp. 55–63.

Liu, Y.-Y. and Bian, J.-S. (2010), "Hydrogen Sulfide Protects Amyoid-ß Induced Cell Toxicity in Microglia", *Journal of Alzheimer's Disease* ,Vol. 22, pp. 1189–1200.

Löffler, Petrides and Heinrich (2007), *Biochemie & Pathobiochemie,* 8th ed., Springer.

Lv, M., Li, Y., Ji, M.H., Zhuang, M., Tang, J.H. (2014), "Inhibition of invasion and epithelial-mesenchymal transition of human breast cancer cells by hydrogen sulfide through decreased phospho-p38 expression", *Mol Med Rep.* , Vol. 10 No. 1, pp. 341-346.

Matsunami, M., Tarui, T., Mitani, K. (2009), "Luminal hydrogen sulfide plays a pronociceptive role in mouse colon", *Gut*, Vol. 58, pp. 751–61.

Moschou, P. N., Roubelakis-Angelakis, K. A. (2014), "Polyamines and programmed cell death" , *Journal of Experimental Botany*, Vol. 65 No. 5, pp. 1285–1296.

O'Dell, T., Hawkins, R.D., Kandel, E.R. and Arancio, O. (1991), "Tests of the roles of two diffusible substances in long-term potentiation: Evidence for nitric oxide as a possible early retrograde messenger", *Proceedings of the National Academy of Sciences No.* 88, pp. 11285–11289.

Olson, K.R., Donald, J.A., Dombkowski, R.A. and Perry, S.F. (2012), "Evolutionary and comparative aspects of nitric oxide, carbon monoxide and hydrogen sulfide", *Respiratory Physiology & Neurobiology*, Vol. 184 No. 2, pp. 117–129.

Pan, Y., Ye, S., Yuan, D., Zhang, J., Bai, Y., Shao, C. (2014), "Hydrogen sulfide (H2S)/cystathionine γ-lyase (CSE) pathway contributes to the proliferation of hepatoma cells" , *Mutat. Res. Fundam. Mol. Mech. Mutagen*, ahead of print, pp. 763-764.

Parekh, A.B. and Penner, R. (1997), "Store Depletion and Calcium Influx", *The American Physiological Society*, pp. 901–930.

Parys, J.B., Bootman, M., Yule, D.I. and Bultynck, G. (2014), *Calcium Techniques: A Laboratory Manual*, CSH Press.

Peers, C., Bauer, C.C., Boyle, J.P., Scragg, J.L. and Dallas, M.L. (2012), "Modulation of ion channels by hydrogen sulfide", *Antioxidants & redox signaling*, Vol. 17 No. 1, pp. 95–105.

Perry, M.M., Hui,C.K., Whiteman, M., Wood, M.E., Adcock, I., Kirkham, P., Michaeloudes, C., Chung, K.F. (2011), "Hydrogen Sulfide Inhibits Proliferation and Release of IL-8 from Human Airway Smooth Muscle Cells", *American Journal of Respiratory Cell and Molecular Biology*, Vol. 45 No. 4, pp. 746-752.

Predmore, B.L., Kondo, K., Bhushan, S., Zlatopolsky, M.A., King, A.L., Aragon, J.P., Grinsfelder, D.B., Condit, M.E. and Lefer, D.J. (2012), "The polysulfide diallyl trisulfide protects the ischemic myocardium by preservation of endogenous hydrogen sulfide and increasing nitric oxide bioavailability", *American journal of physiology. Heart and circulatory physiology*, Vol. 302 No. 11, pp. H2410-8.

Pupo, E., Fiorio Pla, A., Avanzato, D., Moccia, F., Avelino Cruz, J.-E., Tanzi, F., Merlino, A., Mancardi, D. and Munaron, L. (2011), "Hydrogen sulfide promotes calcium signals and migration in tumor-derived endothelial cells", *Free Radical Biology and Medicine*, Vol. 51 No. 9, pp. 1765–1773.

Reiffenstein, R.J., Hulbert, W.C. and Roth, S.H. (1992), "Toxicology of Hydrogen Sulfide", *Annu. Rev. Pharmacol. Toxicl.*

Rosenegger, D., Roth, S. and Lukowiak, K. (2004), "Learning and memory in Lymnaea are negatively altered by acute low-level concentrations of hydrogen sulphide", *The Journal of experimental biology*, Vol. 207 Pt 15, pp. 2621–2630.

Sekiguchi, F., Miyamoto, Y., Kanaoka, D., Ide, H., Yoshida, S., Ohkubo, T., Kawabata, A. (2014), "Endogenous and exogenous hydrogen sulfide facilitates T-type calcium channel currents in Cav3.2-expressing HEK293 cells", *Biochemical and Biophysical Research Communications*, Vol. 445 No. 1, pp. 225-229

Simpson, K. (2011), "Fura and Indo Ratiometric Calcium Indicators", *Molecular Probes Product Information*.

Sitdikova, G.F., Weiger, T.M. and Hermann, A. (2010), "Hydrogen sulfide increases calcium-activated potassium (BK) channel activity of rat pituitary tumor cells", *Pflügers Archiv European journal of physiology*, Vol. 459 No. 3, pp. 389–397.

Stevens, C.F. and Wang, Y. (1993), "Reversal of long-term potentiation by inhibitors of haem oxygenase", *Nature*, Vol. 364 No. 6433, pp. 147–149.

Szabo, C., Coletta, C., Chao, C., Modis, K., Szczesny, B., Papapetropoulos, A. and Hellmich, M.R. (2013), "Tumor-derived hydrogen sulfide, produced by cystathionine- -synthase, stimulates bioenergetics, cell proliferation, and angiogenesis in colon cancer", *Proceedings of the National Academy of Sciences*, Vol. 110 No. 30, pp. 12474–12479.

Szabo, C. and Hellmich, M.R. (2013), "Endogenously produced hydrogen sulfide supports tumor cell growth and proliferation", *Cell Cycle*, Vol. 12 No. 18, pp. 2915–2916.

Tang, G., Wu, L. and Wang, R. (2010), "Interaction of hydrogen sulfide with ion channels", *Clinical and Experimental Pharmacology and Physiology*, Vol. 37 No. 7, pp. 753–763.

Tang, X.Q., Yang, C.T., Chen, J., Yin, W.L., Tian, S.-W., Hu, B., Feng, J.q. and Li, Y.J. (2008), "Effect of Hydrogen Sulphide on Beta-Amyloid-Induced Damage in PC12 Cells", *Clinical and Experimental Pharmacology and Physiology*, Vol. 35, pp. 180-186.

Tang, X.Q., Zhuang, Y.Y., Zhang, P., Fang, H.R., Zhou, C.F., Gu, H.F., Zhang, H. and Wang, C.Y. (2013), "Formaldehyde impairs learning and memory involving the disturbance of hydrogen sulfide generation in the hippocampus of rats", *Journal of molecular neuroscience MN*, Vol. 49 No. 1, pp. 140–149.

Telezhkin, V., Brazier, S.P., Cayzac, S., Müller, C.T., Riccardi, D., Kemp, P.J. (2009), " Hydrogen Sulfide Inhibits Human BK_{Ca} Channels", *Advances in Experimental Medicine and Biology*, Vol. 648, pp. 65-72.

Trevisani, M., Patacchini, R., Nicoletti, P. (2005), "Hydrogen sulfide causes vanilloid receptor 1-mediated neurogenic inflammation in the airways", *British Journal of Pharmacology*, Vol. 145, pp. 1123–31.

Vadivel, A., Alphonse, R.S., Ionescu, L., Machado, D.S., O'Reilly, M., Eaton, F., Haromy, A., Michelakis, E.D., Thébaud, B., "Exogenous hydrogen sulfide (H2S) protects alveolar growth in experimental O2-induced neonatal lung injury" , *PLoS One*, Vol. 9 No. 3, pp. e90965

Wack, N., "Widefield Application Letter. Widefield Calcium Imaging with Calcium Indicator Fura2", *reSolution*, Vol. 2007.

Wang, Z., Liu, D.-X., Wang, F.-W., Zhang, Q., Du, Z.-X., Zhan, J.-M., Yuan, Q.-H., Ling, E.-A. and Hao, A.-J. (2013a), "L-Cysteine promotes the proliferation and differentiation of neural stem cells via the CBS/H2S pathway", *Neuroscience*, Vol. 237, pp. 106–117.

Wang, Z., Zhan, J., Wang, X., Gu, J., Xie, K., Zhang, Q. and Liu, D. (2013b), "Sodium hydrosulfide prevents hypoxia-induced behavioral impairment in neonatal mice", *Brain Research*, Vol. 1538, pp. 126–134.

Weiger, T.M. and Hermann, A. (2014), "Cell proliferation, potassium channels, polyamines and their interactions: a mini review", *Amino Acids*, Vol. 46 No. 3, pp. 681–688.

Weisell, J., Hyvönen, M.T., Alhonen, L., Vepsäläinen, J., Keinänen, T.A., Khomutov, A.R. (2014), "Charge Deficient Analogues of the Natural Polyamines" , *Current Pharmaceutical Design*, Vol. 20, pp. 262-277.

Wen, X., Qi, D., Sun, Y., Huang, X., Zhang, F., Wu, J., Fu, Y., Ma, K., Du, Y., Dong, H., Liu, Y., Liu, H. and Song, Y. (2014), "H2S attenuates cognitive deficits through Akt1/JNK3 signaling pathway in ischemic stroke", *Behavioural brain research*, Vol. 269, pp. 6–14.

Wu, Y.C., Wang, X.J., Yu, L., Chan, F.K.L., Cheng, A.S.L., Yu, J., Sung, J.J.Y., Wu, W.K.K., Cho, C.H. and Deb, S. (2012), "Hydrogen Sulfide Lowers Proliferation and Induces Protective Autophagy in Colon Epithelial Cells", *PLoS ONE*, Vol. 7 No. 5, pp. e37572.

Yamashita, T., Nishimura, K., Saiki, R., Okudaira, H., Tome, M., Higashi, K., Nakamura, M., Terui, Y., Fujiwara, K., Kashiwagi, K. and Igarashi, K. (2013), "Role of polyamines at the G1/S boundary and G2/M phase of the cell cycle", *The International Journal of Biochemistry & Cell Biology*, Vol. 45 No. 6, pp. 1042–1050.

Yang, G., Cao, K., Wu, L. and Wang, R. (2004), "Cystathionine -Lyase Overexpression Inhibits Cell Proliferation via a H2S-dependent Modulation of ERK1/2 Phosphorylation and p21Cip/WAK-1", *Journal of Biological Chemistry*, Vol. 279 No. 47, pp. 49199–49205.

Xu, J., Ji, J. and Yan, X.-H. (2012), "Cross-talk between AMPK and mTOR in regulating energy balance", *Critical reviews in food science and nutrition*, Vol. 52 No. 5, pp. 373–381.

Zhao, W., Zhang, J., Lu, Y., Wang, R. (2001), "The vasorelaxant effect of H2S as a novel endogenous gaseous KATP channel opener", *EMBO* Journal, Vol. 20, pp. 6008–16.

Zhao, Y., Wei, H., Kong, G., Shim, W., Zhang, G. (2013), "Hydrogen sulfide augments the proliferation and survival of human induced pluripotent stem cell-derived mesenchymal stromal cells through inhibition of BK_{Ca}", *Cytotherapy*, Vol. 15, pp. 1395-1405.

Acknowledgements

I am eternally grateful to many people for their continuous support and encouragement which was invaluable for the successful finalization of this work. In the following lines I would like to acknowledge some of them, but I am aware that there were many more who supported me and these words cannot express the gratitude and respect I feel for all of those.

Firstly I would like to thank o. Prof. Anton Hermann and ao. Prof. Thomas M. Weiger who initiated this project and gave me the opportunity to work in their lab. My sincere thanks go to many friends and colleagues for scientific discussion, advice and continuous support always so greatly appreciated, among them Dr. Karin Oberascher-Holzinger who introduced me to the secrets of working in a cell culture lab.

I like to express further greatest thanks for discussing the current literature with me to Prof. Gary Baxter from Cardiff University. Our discussions were always broadening my horizons and helped me immensely in gathering significant literature for this project.

Last but not least I would like to thank my parents for their continuous encouragement and financial support throughout my education. Special thanks goes to my little sister Petra who always had the time to proof read my work and offered constructive criticism.

Only a life lived for others is a life worthwhile.

Albert Einstein

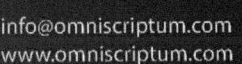